Christopher Mwau Mutungi

Studies on formation and characteristics of enzyme resistant starch from cassava (*Manihot esculenta* Crantz)

TUD*press*

Christopher Mwau Mutungi

Studies on formation and characteristics of enzyme resistant starch from cassava (*Manihot esculenta* Crantz)

TUD*press*

2011

Gedruckt mit Unterstützung des
Deutschen Akademischen Austauschdienstes.

Bibliografische Information der Deutschen Nationalbibliothek
Die Deutsche Nationalbibliothek verzeichnet diese Publikation in der Deutschen Nationalbibliografie; detaillierte bibliografische Daten sind im Internet über http://dnb.d-nb.de abrufbar.

Bibliographic information published by the Deutsche Nationalbibliothek
The Deutsche Nationalbibliothek lists this publication in the Deutsche Nationalbibliografie; detailed bibliographic data are available in the Internet at http://dnb.d-nb.de.

ISBN 978-3-942710-38-1

Verlag der Wissenschaften GmbH
Bergstr. 70 | D-01069 Dresden
Tel.: 0351/47 96 97 20 | Fax: 0351/47 96 08 19
http://www.tudpress.de

Gesetzt vom Autor.
Printed in Germany.

Studies on formation and characteristics of enzyme resistant starch from cassava (*Manihot esculenta* Crantz)

Dissertation submitted for the award of academic grade Doktor-Ingenieur (Dr.-Ing.)

Christopher Mwau Mutungi (B.Sc., M.Sc.)

Faculty of Mechanical Engineering

Institute of Food Technology and Bioprocess Engineering

Technische Universität Dresden

Date submitted for examination	31. 03. 2011
Date of disputation	04. 08. 2011

Referees:

Prof. Dr. Harald Rohm	Technische Universität Dresden, Germany
Prof. Dr. Meinolf Lindhauer	Max Rubner-Institute, Detmold, Germany

Chair of commission:

Prof. Dr. R. Lange	Technische Universität Dresden, Germany

Foreword and acknowledgements

This dissertation is a broad-based discussion of the results that the author obtained during doctorate research at the Institute of Food Technology and Bioprocess Engineering at the Dresden University of Technology, Germany, from April 2008 to March 2011. Specific results have been published in seven original articles and detailed descriptions of the methods are provided in the individual publications.

I acknowledge my supervisor Professor Dr. Harald Rohm without whose support and overwhelming guidance the accomplishment of this dissertation and publication of the seven articles that constitute it would not have been possible. I further acknowledge the willingness of Prof. Dr. Meinolf Lindhauer to serve as external examiner.

My sincere appreciations also go to Dr. Doris Jaros and Dr. Calvin Onyango for their invaluable support and encouragement. I am deeply indebted to Dr. Thomas Doert, Dr. Silvia Paasch, Dr. Steffen Thiele, DLC Kathrin Trabs (Department of Chemistry and Food Chemistry), Dr. Lars Passauer (Institute of Plant and Wood Chemistry), Dipl.-Ing Petra Fiala (Mechanical Process Engineering), as well as Dr. Albena Lederer, and Dipl.-Ing. Susanne Boye (Leibnitz Institute of Polymer Research Dresden), for their assistance with various analytical methods. My further appreciations go to Dipl.-Ing. Felix Rost, and Dipl.-Ing. Stefan Schuldt who undertook parts of the work for their study projects.

I am grateful to the Germany Academic Exchange Service (DAAD) for a scholarship (April 2008 to August 2011) to undertake doctorate research in Germany. My appreciations also go to the collegial team at the working institute for an affectionate collegiality that enabled me to conduct experiments without major hiccups. In a very special way, I acknowledge close acquaintances who made my stay in Dresden an unforgettable experience. My final and special thanks go to my beloved wife, Florence Kanini, and my family for their patience and support during the time of this research.

Vorwort und Danksagung

Die vorliegende Dissertation ist eine breit angelegte Diskussion von Forschungsergebnissen, welche der Autor am Institut für Lebensmittel- und Bioverfahrenstechnik der Technischen Universität Dresden, Deutschland zwischen April 2008 und März 2011 erzielt hat. Spezifische Resultate wurden in sieben wissenschaftlichen Artikeln veröffentlicht, die Methoden in den einzelnen Publikation detailliert beschrieben.

Mein besonderer Dank richtet sich an meinen Doktorvater Herrn Professor Harald Rohm, ohne dessen außergewöhnlicher Unterstützung, Rat und Förderung die Fertigstellung der Dissertation sowie die Veröffentlichung von sieben Publikationen nicht möglich gewesen wäre. Ebenso gilt mein Dank Herrn Professor Meinolf Lindhauer für sein Interesse an der Doktorarbeit und seine Bereitschaft zu Übernahme der Gutachtertätigkeit.

Ausdrücklich möchte ich mich bei Dr. Doris Jaros und Dr. Calvin Onyango für ihre unschätzbare Unterstützung und Förderung bedanken. Im Weiteren danke ich Dr. Thomas Doert, Dr. Silvia Paasch, Dr. Steffen Thiele, DLC Kathrin Trabs (Fachrichtung Chemie und Lebensmittelchemie), Dr. Lars Passauer (Institut für Planzen und Holzchemie), Dipl.-Ing. Petra Fiala (Mechanische Verfahrenstechnik) und Dr. Albena Lederer sowie Dipl.-Ing. Susanne Boye (Leibnitz-Institut für Polymerfoschung Dresden e.V) für ihre Unterstützung bei verschiedenen Analyse-Methoden. Zusätzlich gilt mein Dank Dipl.-Ing. Felix Rost und Dipl.-Ing. Stefan Schuldt, welche Teile der Arbeit innerhalb eigener Studienprojekte voran getrieben haben.

Bedanken möchte ich mich auch beim Deutschen Akademischen Austauschdienst (DAAD) fürs Stipendium zur Durchführung meiner Promotionsforschungen in Deutschland. Weiterer Dank gilt dem herzlichen und kollegialen Team am Institut, welches mir die Durchführung meiner Experimente ohne größeres Stolpern ermöglichte, und meinen nahen Bekannten, die meinen Aufenthalt in Dresden zu einem unvergesslichen Erlebnis machten. Abschließend und im Besonderen danke ich meiner geliebten Frau Florence Kanini und meiner Familie für das Verständnis und die Aufmunterung während des gesamten Studiums.

STUDIES ON FORMATION AND CHARACTERISTICS OF ENZYME RESISTANT STARCH FROM CASSAVA

Mutungi, C., Technische Universität Dresden, Faculty of Mechanical Engineering.

Dissertation 2011, 59 pages, 19 figures, 2 tables, 7 appendices.

Microbiota residing in the human gut is integral to health, and the growth, metabolism, and viability are dependent on the nature of dietary components that escape digestion and absorption in the upper gut. Some forms of starch resist digestion due to structural reasons, and reach the lower gut where fermentation into metabolites that support health occurs. Short chain fatty acids, butyrate in particular, mediate the majority of health benefits. Recrystallised starch is a superior substrate for butyrate because of a structure that probably tilts fermentation pathways in favour of butyrate production.

Delivery of the health-benefits of resistant starch (RS) is possible through fortification of frequently consumed foods. To facilitate enrichment in a wide range of foods, industrial-scale production is enabled through conversion processes that have a high turnover of structurally stable preparations. Possibilities of enzyme mediated synthesis of RS from cassava starch were investigated. A high RS yield of above 80 % was possible by hydrolysis of the α-D-(1 → 6) branching points subsequent to gelatinisation, followed by hydrothermal stabilisation of the linearised polymers.

Structure of RS is important because it influences functionality and interaction with constituents of food systems where the starch is added as functional ingredient. Differential scanning calorimetry, X-ray diffraction, Fourier-transform infrared spectroscopy, Raman spectroscopy, ^{13}C cross polarisation / magic angle spinning nuclear magnetic resonance were used for structural elucidation. Ordered and amorphous phases constitute structure of recrystallised starch; the ordered phase comprises crystalline and non-crystalline double- and single-helical polymers, as well as constrained conformations depending on the recrystallisation method. This structural heterogeneity is responsible for a disparate resistance to digestion by different RS preparations, and differences in techno-functional properties.

UNTERSUCHUNGEN ZUR AUSBILDUNG UND EIGENSCHAFTEN VON ENZYMRESISTENTER MANIOKSTÄRKE

Mutungi, C., Technische Universität Dresden, Fakultät Maschinenwesen.

Dissertation 2011, 59 Seiten, 19 Abbildungen, 2 Tabellen, 7 Anhänge.

Eine integrale Vorraussetzung für die menschliche Gesundheit ist die Darmflora. Deren Wachstum, Metabolismus und ihre Viabilität werden maßgeblich von der Art und Zuführung von Ballaststoffen beeinflusst, welche sich der Digestion und Absorption im oberen Darm entziehen. Einige Arten von Stärke widerstehen der Verdauung aufgrund struktureller Eigenarten und werden im unteren Darm zu gesundheitsfördernden Metaboliten fermentiert. Rekristallisierte Stärke gilt als Substrat für die Bildung von Butyrat und kurzkettige Fettsäuren, deren gesundheitsfördernde Eigenschaften eine ausgewogene Darmflora fördern.

Die Gewährleistung der gesundheitsfördernden Wirkung von resistenter Stärke (RS) kann durch eine Anreicherung in regelmäßig konsumierten Nahrungsmitteln ermöglicht werden. Im Rahmen der Arbeit wurde die Möglichkeit einer enzyminduzierten Synthese von RS aus Maniok untersucht. Dabei konnte durch Hydrolyse der α-D-(1 $\rightarrow$ 6)-Verzweigungen nach vorangegangener Verkleisterung und anschließender hydrothermischer Stabilisierung der linearisierten Polymere eine RS-Ausbeute von über 80 % erzielt werden.

DSC, XRD, FTIR, Raman-Spektroskopie, ^{13}C CP/MAS NMR und dynamische Wasserdampfsorption wurden zur strukturellen Bewertung der Stärkepräparate herangezogen. Geordnete und amorphe Bereiche bestimmen die Struktur von rekristallisierter Stärke; die geordneten Regionen beinhalten kristalline und nicht-kristalline doppel- und einfach-helicale Anordnungen, in Abhängigkeit der Art der Rekristallisation aber auch erzwungene Konformationen. Diese strukturelle Heterogenität ist als Ursache für variable Verdauungsresistenzen und techno-funktionelle Eigenschaften durch verschiedene RS-Präparationen zu sehen.

Table of contents

1. Introduction

Early epidemiological studies on dietary fibre research indicate that populations that consumed high proportions of non-starch polysaccharides (NSP) in daily diets suffered less from gastrointestinal diseases as compared to populations that thrived on diets rich in fat and protein, but low in NSP (Burkitt, 1973). In this respect, increasing the amount of dietary NSP was suggested to contribute to reduction of the risk for gastrointestinal disorders. Later studies, however, have revealed that indigestible dietary components have a more positive impact on health and illness-prevention if they can be fermented by gut microbiota during gut transit (Topping et al., 2008). NSP brings about gastrointestinal protection through physical mechanisms such as increased faecal bulking, dilution of toxic bowel contents and speeding up of gut transit. However, the regular consumption of highly fermentable dietary fibre introduces additional effects that include gut associated immune modulation, balancing of colonic microbiota, and increased production of health promoting metabolites, among them, the short chain fatty acids (SCFAs) butyrate, propionate and acetate. Butyrate is a prime substrate for colonocyte energy metabolism, and participates in processes that lower risk of gut inflammation and colorectal cancer by enhancing normal colonocyte phenotype and healthy gut epithelium, whereas propionate participates in stabilisation of blood glucose and holds back cholesterol synthesis in the liver (Hamer et al., 2008; Lim et al., 2005; Topping & Clifton, 2001).

Starch that resists digestion (Resistant starch, RS) is a major substrate for gut bacteria that produce SCFAs (Bird et al., 2000; Topping et al., 2001). If fermentation of the starch in the colon is selective, for example, in favour of health-promoting bacteria, a sustained modulation of gut microbiota could also be facilitated through mechanisms involving substrate stimulated metabolism, growth, and viability of the specific probiotic groups (Bird et al., 2000; Crittenden et al., 2001; Nobakhti et al., 2009; Topping et al., 2003). Moreover, an immediate impact of RS ingestion is a lowered glycemic load and post-ingestion glycemia, conditions that are relevant in lowering the risk for type II diabetes (Raben et al., 1994; Reader et al., 1997).

Generally, RS comprises starch that escapes action of digestive enzymes, and any products of the digestion that may bypass absorption in the small intestine of healthy individuals (Asp, 1992). Starch may escape digestion if it is physically locked within indigestible cellular matrices (RS I), is ungelatinised and therefore inaccessible in the compact native granules (RS II), is gelatinised but then reassembled into physical structures that deter amylolysis (RS III), or is chemically modified so that bulky derivatisation results in sterical hindrance to enzyme-substrate complex formation (RS IV) (Sajilata et al., 2006). RS I and RS II are essentially lost during food processing and preparation, and are therefore of limited technological significance (Haralampu, 2000), whereas RS IV is less popular because of the heavy chemical load. RS III, also known as recrystallised starch, withstands food processing conditions, and its properties may be modified through physical or enzymatic transformations to suit specific technological and functional needs. These attributes make RS III a carbohydrate of interest to food technologists and nutritionists as a functional food ingredient.

Industrial-scale production of RS is one way to make fortification of foods with the functional carbohydrate possible. A practical option is through enhancing recrystallisation of processed starch. Various investigators reported different methods for RS III production (Sajilata et al., 2006), but to enable enrichment in a wide range of foods, methods leading to high RS III yield are necessary. This way, the functional starch can be made available as a ready-to-use ingredient for frequently consumed foods or as supplementation nutraceutical product. Where inclusion in composite recipes and further processing is intended, preparations should be able to tolerate a wide range of conditions that characterise the diverse food processing and preparation techniques of the modern food industry, as well as sustain or achieve superior technological attributes specific to these foods. Furthermore, some investigations have, quite interestingly, shown that the physical and molecular structure of RS have influence on the interaction with gut microbiota (Bird et al., 2000). These needs inspired the current investigations on methods to increase RS III formation, and the characterisation of the structure of different preparations.

The motivation to use starch from cassava (*Manihot esculenta Crantz*) (Figure 1) was the need to add value to this excellent starch source, whose utilisation at both domestic and industrial levels is still low despite its wide-spread cultivation in the tropics. In parts of sub-Saharan Africa, Asia and Latin America, cassava is grown as a primary or strategic famine food crop, and is traditionally consumed as raw, boiled, or fried chunks, or as flour constituted into thin or thick porridge either alone or in combination with other flours, particularly of cereal origin. The main limitation to cassava utilisation is perishability. Freshly harvested cassava root tubers undergo rapid post-harvest physiological deterioration within 2 to 3 days after harvest. As a preservation and value addition measure, the starch, from fresh roots or dried chips, may be extracted for use in foods, paper, textile, adhesives and fuel industries. Extraction of starch from cassava is relatively easy, and involves a simple wet milling process. Key steps entail peeling and washing of the roots, rasping in excess water, filtering, washing of the starch, and finally drying (Figure 1). About 25 to 30 % starch may be obtained from mature, good quality tubers, and approximately 60 % can be extracted from dried cassava chips. The starch is usually low in impurities such as proteins and lipids, whose contents are below 1.7 % and 0.4 % on dry matter basis, respectively (Charles et al., 2005). A relatively low popularity of cassava as food is also because of its low nutritional versatility which puts it on unfavourable competition course with other more stable and nutritionally resourceful energy sources such as corn, wheat and rice. These reasons motivate investigations on diversification of products that can be derived from cassava, especially in a manner that extends beyond the traditional utilisation options.

Figure 1: Cassava growing in the field (a), root tubers of the uprooted plant (b), transverse section of a fresh root tuber (c), and small scale extraction of starch from fresh cassava roots: pealing and washing (d), rasping in excess water (e), separation of fibrous material (f), sedimentation of the starch milk (g), and sun-drying of the starch cake after decantation (h). Part (i) shows appearance of the native starch granules under scanning electron microscope.

2. General discussion

2.1. Impact of resistant starch on health

Gut microbiota is an integral part of human health (Sekirov et al., 2010), and the growth and metabolism depends on dietary ingredients that escape digestion and absorption in the upper gut. These indigestible dietary components are fermented in the colon into organic acids, gases and biomass (Louis et al., 2007; Macfarlane et al., 2008). Among the organic acids SCFAs, namely acetate, propionate and butyrate, and to a lesser extent, formate, valerate, caproate and their iso-forms, are formed as major end-products (Macfarlane & Macfarlane, 2003). Lactate and succinate are also formed in significant amounts, but these do not accumulate in the colon of healthy individuals because they are utilised in metabolic cross-feeding interactions among colonic bacteria (Louis et al., 2007; Macfarlane & Macfarlane, 1993). The gases, (hydrogen, carbon dioxide and methane) are largely expelled in breath or as flatus, although small amounts of carbon dioxide and hydrogen also end up in certain metabolic pathways (Flint et al., 2007).

SCFAs mediate the majority of health benefits associated with fermentable non-digestible dietary components (Figure 2). A particularly important SCFA is butyrate because of its role in combating gastrointestinal inflammation and carcinogenesis (Hamer et al., 2008). In comparison with other indigestible ingredients such as prebiotic fructo-oligosacharides, resistant starch is more aggressively metabolised into butyrate (Bird et al., 2000; Louis et al., 2007; Topping et al., 2001; Topping et al., 2003). However, the butyrogenic effects of different types of resistant starch are seemingly disparate with regard to the absolute amounts, the relative proportions with respect to other SCFAs, and the distribution of butyrate along the colon (Brouns et al., 2007; Fassler et al., 2006a; Ferguson et al., 2000; Goni et al., 2000; Lehmann et al., 2002; Martin et al., 2000; Reid et al., 1998; Rideout et al., 2008; Zampa et al., 2004). RS III has superior ability to yield butyrate than the other RS types, and in some instances, also exhibits lactobacillus and bifidobacteria enhancing effects (Bouhnik et al., 2004; Brouns et al., 2007; Brown et al., 1998; Schwiertz et al., 2002; Silvi et al., 1999; Wronkowska et al., 2008; Zampa et al., 2004).

Numerous factors related to hosts colonic microbiota and the substrate itself, influence the metabolism of non-digestible substrates into SCFAs. Host microbiota factors include the relative numbers, substrate specificities or preferences, and fermentation pathways of individual bacterial groups among other more specific considerations such as availability of inorganic terminal electron acceptors and the need for disposal of reducing equivalents arising from substrate utilisation (Cummings et al., 1996; Macfarlane & Gibson, 1997; Macfarlane et al., 1993; Pryde et al., 2002; Rabiu & Gibson, 2002; Zampa et al., 2004). Butyrate and propionate, for example, are reduced products, and their formation is predisposed to activities of hydrogen-utilising bacteria. For this reason, butyrate and propionate are produced by a restricted diversity of bacterial groups, whereas acetate is generated by most heterotrophic gut anaerobes through reductive acetogenesis (Louis & Flint, 2009; Miller & Wolin, 1996; Pryde et al., 2002).

Substrate specificity is seen in the ability of some SCFA-producing bacteria, for example *Eubacterium rectale* and *Roseburia* spp that produce butyrate from starch, to fundamentally degrade the substrates. A host of other SCFA producers are unable to degrade the primary substrates and depend on degraded residues in what is termed as metabolic cross-feeding. Furthermore, stimulation of production of some SCFAs can occur indirectly. Acetate and lactate, for instance, act as precursors for butyrate (Belenguer et al., 2006; Duncan et al., 2004a; Duncan et al., 2004b; Morrison et al., 2006; Pryde et al., 2002). The implication is that enhanced activity of acetate or lactate producing microbiota such as lactic-acid bacteria in the proximal colon could synergistically boost butyrate formation. Other factors related to substrate (indigestible/ unabsorbed dietary residues) include the amount, type, degree of processing, degree of polymerisation, bonding structure and nature of matrices within which the components are sheltered. These aspects also relate to transit dynamics in the colon. A further significant consideration is the substrate's chemical composition and redox potential. Highly reducing starch substrates may tilt the fermentative metabolism of individual bacteria, for example, in favour of butyrate, because of its role as a hydrogen sink (Pryde et al., 2002).

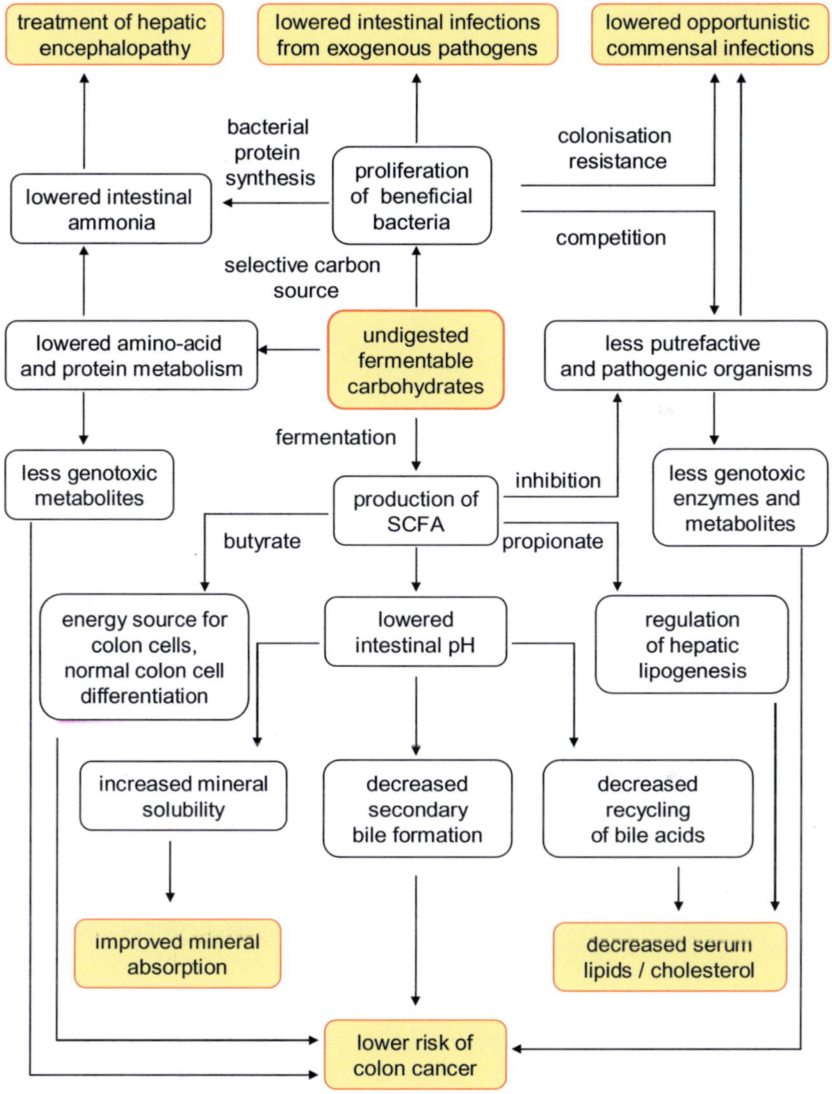

Figure 2: Mechanisms for improved health due to dietary intake of indigestible fermentable carbohydrates (Crittenden, 1999).

The superior butyrogenic property of RS III as compared to other forms of RS might as well be related to structural properties, where by, moderately stable RS III preparations achieve butyrogenicity in two probable ways: first by way of the less stable fraction being able to undergo substantial fermentation in the proximal colon so that products such as lactate and acetate enter the distal colon as ready precursors for butyrate, and secondly, by way of significant amounts of the more stable fraction (together with other simpler starch degradation residues), reaching the distal colon where butyrate fermentation is more relevant (Dongowski et al., 2005; Jacobasch et al., 2006; Martin et al., 1998; Martin et al., 2000). From this point of view, structural properties of RS III may be considered useful markers for developing preparations that have specific functionality, as they not only influence the specificity of interaction with intestinal microflora, but also determine the nature, site and intensity of the physiological effects. Structural features of RS III that influence enzyme attack such as specific surface, micro-porosity, crystallinity and chain length may be manipulated through processing methods (Dongowski et al., 2005; Lesmes et al., 2008; Robertson et al., 2001). High crystallinity, for example, and a high proportion of A-type crystalline structure might delay microbial invasion enabling the starch to reach the hindgut (Jacobasch et al., 2006). B-type structures, on the other hand, might support lactobacillogenic and bifidogenic effects in addition to butyrogenic effects as well (Brown et al., 1997; Lesmes et al., 2008).

2.2. Structure of native starch

Native starch is deposited as granules in a wide array of tissues in diverse plant species. The granules vary in size (approximate diameter 1-100 µm), shape (round, lenticular, polygonal), size distribution (uni- or bi-modal), granular association (simple or compound) and composition (α-glucan, lipid, moisture, protein and mineral content) depending on botanical origin. Amylose and amylopectin are the main macromolecular components representing 98 to 99 % of the dry granules. Amylose content of normal starches is 20 to 35 %, waxy starches < 15 %, and high amylose starches > 40 % (Tester et al., 2004).

Amylose is a predominantly linear molecule of α-D-(1 → 4) linked glycopyranosyl units with approximately 1 % α-D-(1 → 6) branch linkages (Figure 3). The degree of polymerisation (DPn) ranges between 300-5000 glucose monomers, and a single molecule comprises approximately 3-11 chains, each with an average chain length of about 200-700 glucose units (Buleon et al., 1998; Tester et al., 2004).

Individual amylose chains are randomly located in the amorphous growth rings of the starch granule and, occasionally, may be sandwiched between amylopectin clusters (Figure 4). The number of amylose chains increases towards the surface of the granule, although the size of molecules nearest to the surface is generally smaller than that of interiorly located ones. Surface amylose is capable of leaching out of the granule whereas interior amylose is not, partly because of the larger molecular size, and the interaction that this interior fraction occasionally exhibits with amylopectin (Imberty et al., 1991). Some amylose chains also form inclusion complexes with endogenous lipids.

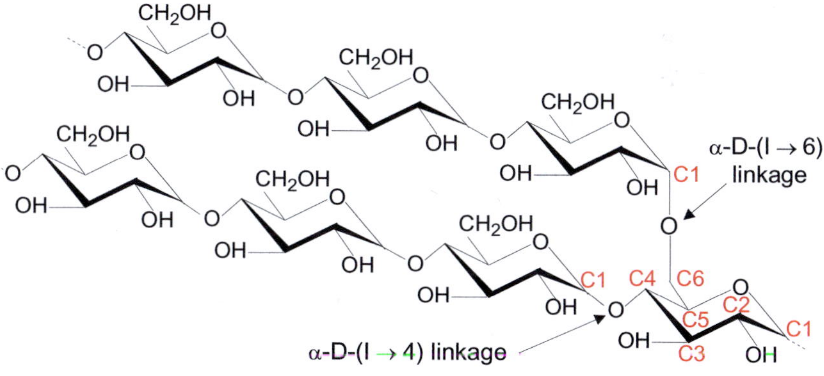

Figure 3: Glycopyranose unit linkage in starch

Amylopectin, on the other hand, is highly branched and comprises 5-6% α-D-(1 → 6) glycosyl linkages (Buleon et al., 2007). A DPn of 9600-15900 is typical, but unlike amylose, the branching pattern is characterised by relatively short side-chains whose average DPn is approximately 18-25 glucose units. Depending on the length and position in the amylopectin molecule, individual amylopectin chains are classified into A- and B-chains (Hizukuri, 1986). The A-chains are the most exteriorly situated chains, and attach to the longer B-chains by α-D-(1 → 6) linkage. The B-chains, in turn, link in a similar manner to the only C-chain that is the 'backbone' of the molecule (Figure 4b). B-chains are further categorised into B_1, B_2, B_3 and B_4 chains based on the length and number of clusters traversed within a single amylopectin molecule. Typical chain lengths of cassava and some other common starches are presented in Table 1.

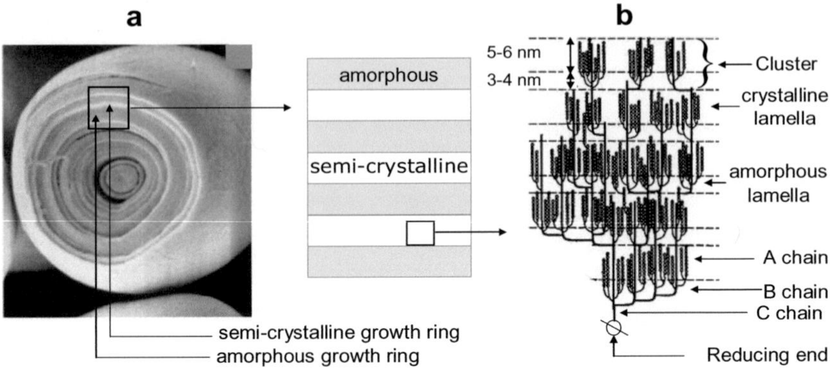

Figure 4: Starch granule ultra-structure showing alternation of amorphous and semi-crystalline growth rings (a) and cluster model of amylopectin (b). The dashed horizontal lines in (b) demarcate amorphous and crystalline lamellae that correspond to branching zones, and crystalline A- and B_1-chain zones, respectively, within individual clusters (Buleon et al., 2007).

Table 1: Molecular properties of cassava and some common normal starches

	Cassava	Potato	Wheat	Maize	Rice
Amylose (%)	19-24	25-26	25-29	25-28	20-28
β-amylolysis limit (%)	75-80	80	79-85	82-84	73-84
DPn ($x10^3$)	2.05-4.39	4.92	0.83-1.57	1.0	1.11
Number of chains	4.7-7.8	7.3	4.8-6.2	3.1-4.4	2.5-4.3
Average chain length	410-550	670	270	305	250-370
Unbranched amylose (% mol/mol)	58-66	-	56-73	52	51-75
Amylopectin					
β-amylolysis limit (%)	57	55	56-59	60	56-59
DPn ($x10^3$)	13.4	9.8	4.8-9.4	8.2-10.2	4.7-15
Average chain length	20-21	22-24	20-21	22	20-22
Component chains					
A	9-12	16	10-13	13	12-13
B_1	21-24	24	20-23	18	16-17
B_2	42	48	41-43	47	41-44
B_3	69	75	68-91	136	85-180
B_4	115	104	-	-	-
Double-helical amylopectin (%)	44	40-64	32-46	38-43	49-63

(Charoenkul et al., 2006; Hoover, 2001; Ramesh et al., 1999; Shibanuma et al., 1994; Tester et al., 2004).

Because only the A- and B_1-chains form double helices, starch granules are semi-crystalline in the native form, and display an inner architecture characterised by 'growth rings' that correspond to concentric semi-crystalline shells separated by amorphous regions (Buleon et al., 1998). The semi-crystalline shells arise from organisation of the amylopectin clusters into alternating amorphous and crystalline lamellae. Amorphous lamellae comprise the branching point zones, whereas the crystalline lamellae are the three-dimensionally packed double helices emanating from adjacent A- and B_1-chains (Figure 4).

2.3. General aspects of starch recrystallisation

Native starch losses the semi-crystalline structure upon heating in excess water. In the order-disorder transition, starch granules absorb water, swell and rupture, enabling the amylose to leach out as a random coil polymer (Jenkins & Donald, 1997). The amorphous starch is, however, capable of re-associating by way of the individual chains forming double helical filaments that subsequently assemble into partially ordered networks or aggregates that are stabilised by hydrogen bonds. The recrystallisation proceeds in three steps: nucleation, propagation and maturation, which also govern the rate and nature of the re-association (Eerlingen & Delcour, 1995).

Nucleation begins when a change in thermodynamic state of the starch gel occurs as a result of cooling or super-saturation, for example, through evaporation or addition of precipitant (Buleon et al., 2007; Long et al., 1995). This kinetic step occurs at any temperature above the glass transition temperature of the amorphous starch, but the rate decreases exponentially with increasing temperature. The mechanisms of nucleation are mainly orientational, whereby starch chains align at surfaces of foreign phases such as water molecules or other impurities (Jagannath et al., 2001; Long et al., 1995). Propagation, on the other hand, entails chain-folding and crystal perfection as the main features, and the rate increases exponentially as the temperature increases further above the glass transition temperature (Eerlingen et al., 1993; Long et al., 1995). Both nucleation and propagation are mobility processes and, for this reason, recrystallisation is optimised when the conditions are chosen such that free energy of the aqueous starch generates sufficient internal forces to enable shuttling of starch chains from disordered to ordered phase, as well as their orientation upon each other once the critical nuclei size has been attained.

Double helical chains dominate the crystalline domains but loose chains, free chain ends and single helical chains may also become entangled, and these manifest as amorphous zones within the crystalline network (Gidley et al., 1995). A schematic model is shown in Figure 5. The extent to which recrystallisation proceeds, and the type of morphologies formed depends on a number of factors such as temperature, starch type (degree of polymerisation and branching structure), concentration of starch in the aqueous system, pH and presence of surface active substances, lipids, electrolytes, proteins and sugars, which in essence influence the kinetics of nucleation and propagation (Eerlingen et al., 1995). Interaction of starch with protein for instance reduces chain mobility, and ions such as calcium and potassium inhibit hydrogen bond formation (Escarpa et al., 1997). Soluble sugars may act as anti-plasticisers (Eerlingen et al., 1994b; Escarpa et al., 1997), whereas lipids and emulsifiers compete with the linear chains for complementary strands (Eerlingen et al., 1994a).

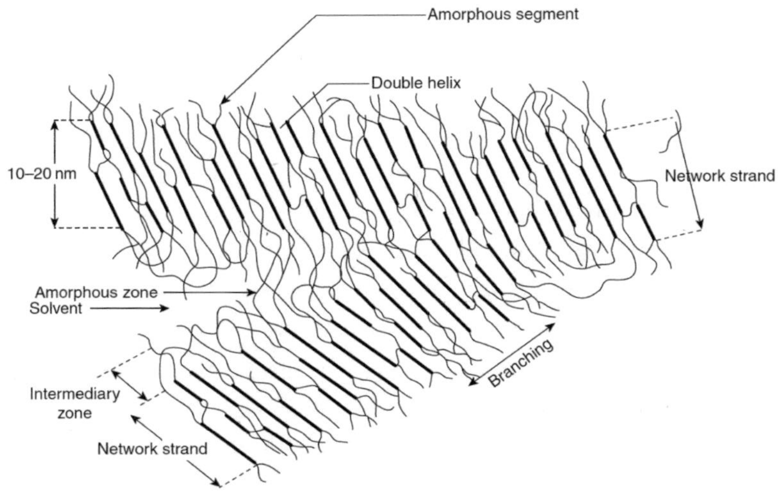

Figure 5: Schematic representation of starch gel network formation. Ordering into crystalline domains occurs over the starch regions comprising double helixes, whereas the free chain ends become entangled within the crystalline domains creating amorphous zone imperfections (Leloup et al., 1992).

2.4. Application of debranching enzymes

A main factor influencing the rate and nature of starch recrystallisation is the molecular structure, that is, the average length and distribution of polymer chains, and degree of branching. One way to enhance re-association is to lower the entropic and conformational constraints to chain ordering by application of a controlled depolymerisation step. Hydrolysis with α-amylase produces shorter-chain starch hydrolysate which exhibits higher tendency to re-associate as compared to the non-hydrolysed gel (Cairns et al., 1990; Htoon et al., 2009). Nevertheless, the starch retains a branched structure, and this impedes stable intra-molecular hydrogen bonding. A further limitation of the α-amylolysis approach is that optimisation of a double-helical structure based on optimal-length chains, is difficult to achieve. Optimal-length chains should not only coincide with absence of major intervening amorphous regions but also be above the minimum degree of polymerisation required for double-helix formation (Gidley et al., 1995). Because of these reasons, yields of resistant starch are relatively low when gels are depolymerised with α-amylase (Onyango & Mutungi, 2008; Appendix 1).

Application of debranching enzymes increases the fraction of optimal chains, as well as the ability of the chains to align into ordered domains. Starch debranching enzymes cleave α-D-(1 → 6) glycodic bonds by an endo-mechanism, releasing linear gluco-oligosaccharides with α-configuration in the anomeric carbon of the reducing end glucose residue. Two types of debranching enzymes are distinguished: pullulanase (Pullulan 6-α-glucanohydrolase, EC 3.2.1.41) and isoamylase (Glycogen 6-α-D-glucanohydrolase, EC 3.2.1.68). Some enzymes categorised under pullulanase also hydrolyse α-D-(1 → 4) bonds in addition to α-D-(1 → 6) bonds. For this reason, pullulanases are further distinguished into pullulanase type I and pullulanase type II. The latter is also called amylopullulanase, because of the α-D-(1 → 4) activity (Bertoldo & Antranikian, 2002). A special type of pullulanase referred to as neopullulanase also mediates transglycosylation reactions, whereby new α-D-(1 → 4) or α-D-(1 → 6) glycosidic bonds are formed (Takata et al., 1992).

During debranching linear starch chains re-associate quickly (Figure 6) and eventually precipitate into polycrystalline aggregates comprising lamellar, spherulitic, and fibrillar morphologies (Figure 7). Formation of these morphologies is governed by the chain length, starch concentration, and rate of cooling (Buleon et al., 2007).

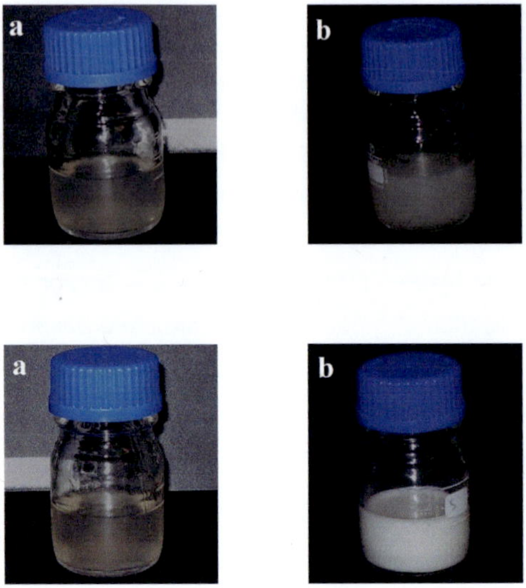

Figure 6: Optical properties of 10 % w/w cassava starch gels at time zero (a) and after 24 h incubation at 50 °C (b). In the upper row, starch gel was mixed with a solution of denatured pullulanase whereas in the lower row, an active form of pullulanase (25 U/g starch) was added before incubation. Turbidity and gel stiffness increase as a result of gel network formation, and is highly enhanced when the starch is mixed with debranching enzyme.

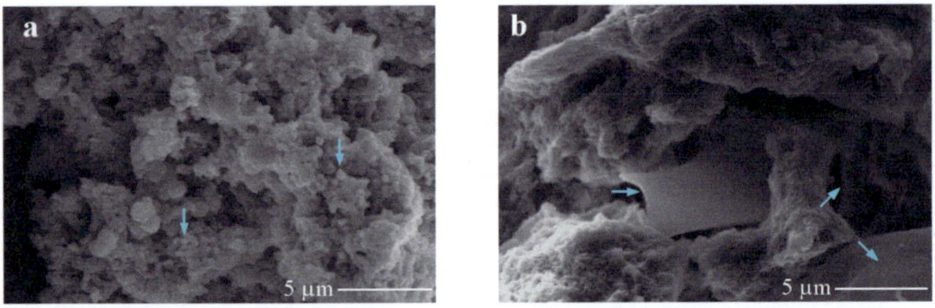

Figure 7: Scanning electron micrographs showing morphological features of debranched freeze-dried cassava starch: (a) minute spherulitic bodies (approximately 200 nm), and (b) plate-like structures (lamellae) inside a fibrillar mass. In (b), native starch was modified with 0.14 mol/L hydrochloric acid at 40°C for 4 days prior to debranching. Lamellar and spherulitic morphologies are formed by the shorter chains where as fibrillar morphologies originate from the longer chains (Buleon et al., 2007).

Preferentially, type I pullulanase debranches pullulan, a polymer comprising α-D-(1 → 6) linked maltotriose chains. It also acts on amylopectin but is incapable of debranching glycogen, which is more vastly branched. Conversely, isoamylase hydrolyses α-D-(1 → 6) linkages in amylopectin and glycogen at rates estimated to be 7 and 124 times faster than pullulanase, respectively (Nakamura, 1996). The rate of hydrolysis of the short branched chains is, however, remarkably slower because isoamylolysis preferentially begins at the longer interior chains (Robyt, 2009). Moreover, whereas pullulanase hydrolyses side chains comprising two or more glucose units, isoamylase can only hydrolyse branched chains exceeding three glucose units. These activity and specificity dissimilarities have two implications: First, they reflect differences in the extent to which starch debranching may be achieved by the two enzymes under similar conditions, involving for example starch concentration or gel viscosity, enzyme concentration, and reaction time (Mutungi et al., 2009a; Appendix 2). Secondly, the kinetics of starch debranching have influence on the structure of the resultant hydrolysate, and therefore the rate and extend of chain re-association (amount of enzyme-resistant starch formed) during the course of debranching (Figure 8 and Figure 9). The presence of contaminating activity in pullulanase, for example, causes lower RS III yield when a high amount of the enzyme solution is used (Figure 9) because of a disproportionately high α-D-(1 → 4) hydrolysis, which brings about shortening of the linear chains, and accumulation of low molecular size oligosaccharides and sugars that impede chain re-association (Mutungi et al., 2009a; Appendix 2).

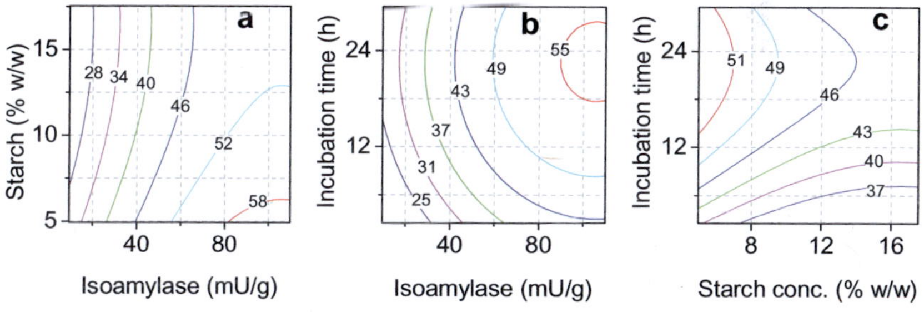

Figure 8: Influence of starch and isoamylase concentrations [incubation time = 16 h] (a), incubation time and isoamylase concentration [starch concentration = 11.3 % w/w] (b), and starch concentration and incubation time [isoamylase concentration 60 mU/g starch] (c), on RS III yield (g/100g) of isoamylase debranched cassava starch gels.

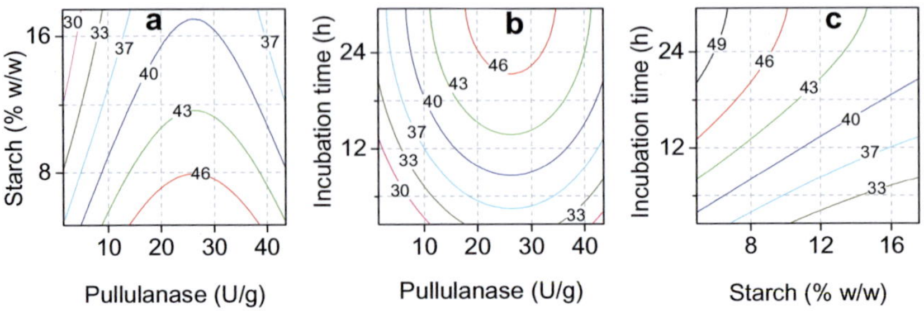

Figure 9: Influence of starch and pullulanase concentrations [incubation time = 16 h] (a), incubation time and pullulanase concentration [starch concentration = 11.3 % w/w] (b), and starch concentration and incubation time [pullulanase concentration = 22.5 U/g] (c), on RS III yield (g/100g) of pullulanase debranched cassava starch gels.

Indigenous amylose in starch also exerts significant effects on association of the short glucan chains in a number of ways. During the course of debranching, these long polymers retrograde rapidly into semi-amorphous networks on which multiple crystalline domains arising from association of shorter linearised amylopectin chains may be assembled. This way, amylose provides a template effect by forming seed nuclei. Nevertheless, aggregation of these multiple domains is hindered by the partially cross-linked amylose networks and, as a consequence, loosely packed morphologies that exhibit low resistance to enzyme hydrolysis are formed (Mutungi et al., 2009b; Appendix 3). Domain association during debranching can be enhanced if a prior step to depolymerise the long amylose polymers is incorporated. However, as already pointed out, application of α-amylase is unattractive, and an alternative approach would be mild reaction of the starch with a low concentration of a mineral acid, such as hydrochloric acid prior to gelatinisation (Mutungi et al., 2010, Appendix 4). Acid reaction preferentially attacks the long amylose chains located in the amorphous growth rings of native starch granules. It also hydrolyses substantially the branching points associated with B-chains of amylopectin in the semi-crystalline lamellae. The reaction results in a narrower chain-length distribution and increased double-helical content.

2.5. Enhancement of enzyme – resistant starch by hydrothermal treatments

Crystalline morphologies that assemble during debranching of aqueous starch suspensions exist as weak metastable forms, and so treatments that enhance plasticisation and interactions between the starch molecules may be applied to modify the stability as well as the techno-functional properties. Hydrothermal treatment approaches are popularly used to modify physicochemical properties of native starches. The treatments may be distinguished into annealing, heat-moisture treatment and temperature-cycling, whereby the defining differences are temperature and moisture conditions (Gunaratne & Hoover, 2002; Silverio et al., 2000; Tester & Debon, 2000; Vermeylen et al., 2006).

Annealing is conducted at a temperature below the melting temperature of any existing crystallites, and in an intermediate or excess moisture content, usually above 40 % w/w (Jacobs & Delcour, 1998). The treatment increases segmental mobility within the amorphous and semi-crystalline domains, resulting in enhanced structural ordering. Specific effects might include modification of crystalline domains to more energetically stable ones, conversion of amorphous chains into double-helical strands, lengthening of double helices as a consequence of progressive helicalisation of adjacent chain ends, and perfection of crystalline structure as a result of enlargement, change of shape or change in direction of their growth. An enhancement of interaction between individual crystallites might also occur (Jayakody & Hoover, 2008).

Heat-moisture-treatment (HMT), on the contrary, relies on high temperature and moisture contents below 40 % w/w (Franco et al., 1995; Gunaratne et al., 2002). Chain mobility is quite subtle during HMT due to the low moisture involved. However, change in chain packing arrangement from the less stable B-form to the more stable A-form is probably the main notable effect. A minimum water content of approximately 20 % is, nevertheless, required to facilitate the structural transition. In this context, crystalline transition during HMT might be considered more or less a solid-state transition, involving progressive removal of inter-helical water and a unidirectional slipping of the chain duplexes (Perez et al., 1990). V-type structures also develop as a consequence of improved packing of single helices (Buleon et al., 2007)., In debranched cassava starch, these conversions are significantly influenced by chain length, and depolymerisation of the long amylose chains, for instance, by controlled acid treatment of the starch prior to debranching favours conversion into A-type crystalline structure (Mutungi et al.,

2010; appendix 4).

In temperature-cycling, starch suspensions, usually containing excess water, are exposed to alternate heating and cooling regimes. The temperature gap between the high and low holding temperatures, the number of alternate exposures and the duration of each cycle determine the alterations of structural properties, namely crystallinity, crystalline type and thermal stability.

2.6. Crystallinity

Starch crystallinity corresponds to long-range ordering, that is, lattice periodicity of helical chains when the polymers aggregate. Wide angle X-ray diffraction is a common probe for long-range order, and reflects molecular structure based on elastic scattering of X-rays by electron clouds of individual atoms. On account of the diffraction pattern generated when the starch is focused with X-rays, native starches exhibit primarily A- and B-type crystalline arrangements (Popov et al., 2009; Takahashi et al., 2004; Wu & Sarko, 1978a; Wu & Sarko, 1978b), which originate from parallel stranded α-D-(1 → 4) glucan double helices. These two forms are identical with respect to double-helical structure; both arrangements comprise counter-clockwise double helices with a helix pitch length of 2.1 nm, equivalent to six glucose units (Gidley & Bulpin, 1987; Imberty et al., 1991). Chain packing is, however, denser in A- than in B-polymorph, so that double helices in A-type crystallites assemble to form a monoclinic unit cell with less inter-helical water, whereas in B-type crystallites, a hexagonal unit cell with more water molecules occupying the central core is created. A- and B- polymorphs are thus alternate hydrate forms having similar chain conformations (Figure 10).

In native starch, A-type crystallinity is characteristic of cereal and waxy starches because the chains that participate in double-helix formation (A and B_1 amylopectin chains) are shorter in these starches. Conversely, B-type structure is predominant in tuber and amylose-rich starches because the double helix forming chains are longer (Cheetham & Tao, 1998; Gidley, 1987; Whittam et al., 1990). In cases where both A- and B-arrangements co-exist, such as in legume and some tuber starches, the packing structure has been classified as C-type. In other instances, amylose inclusion complexes with lipids occur as ordered domains, and the structure has been categorised to as V-type crystallinity (Buleon et al., 1998).

V-structures also assemble when long amylose chains form inclusion complexes with alcohols or flavour compounds, or when single-helical glucans simply subsist as stable, parallel-ordered strands by "nesting" effect (Godet et al., 1995; Le Bail et al., 2005). With regard to chain conformation, V-structures are generally counter-clockwise single helices of six glucose residues per turn for complexes with aliphatic alcohols and monoacyl lipids. However, where guest molecules bulkier than a hydrocarbon chain are involved, helices of seven or eight glucose residues per turn are also possible (Snape et al., 1998).

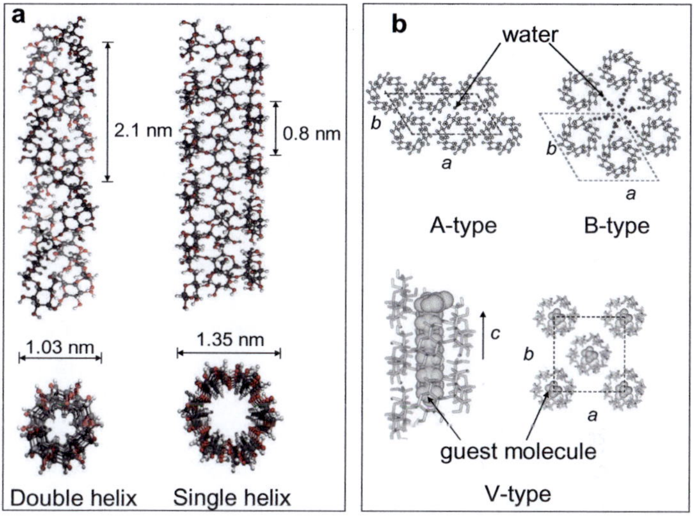

Figure 10: Structure of double- and single-helical glucan chains [bottom illustration is top view] (a), and the packing arrangement of A- B- and V-structural polymorphs (b). A-polymorph is based on monoclinic unit cell packing of double helices with less water molecules between them, B-polymorph is based on hexagonal unit cell packing of double helices with more water molecules in the central core, and V-polymorph is based on single helical structures with or without inclusion molecules, and the packing (on *a, b* plane) might vary depending on the type of the guest molecule (Buleon et al., 1998; Therien-Aubin & Zhu, 2009).

Similar to native starch, recrystallised starch may exhibit A-, B-, C- and V-polymorphic forms. The B-form is kinetically preferred to the A-structure and assembles readily from aqueous starch suspensions as a consequence of rapid chain association (Buleon et al., 2007). Conversely, the A-form, which is thermodynamically more stable, assembles slowly although high recrystallisation temperature, high starch concentration, and presence of dehydrants such as alcohol may favour its formation (Gidley, 1987; Shamai et al., 2003). Moreover, short α-D-(1 → 4) glucans favour recrystallisation into the A-form, whereas longer ones, errand formation of B-crystalline form. A transition from B- to A-polymorph is possible under some conditions such as heat-moisture treatment that induces either helical sliding within crystalline domains or dehydration, and subsequently, relocation of double helix pairs (Mutungi et al., 2009b; Appendix 3).

Significant amounts of V-type structures form when recrystallisation is conducted at high temperature (Shamai et al., 2003). In numerous instances, however, aliphatic lipid chains within amylose helices are rather locked in a solid-like state, that is, they settle in a fixed conformation because orientational adjustment is averted due to steric constraints. Under these circumstances, V-type X-ray diffraction peaks might not be observed, that is, the V-structures are not detected as crystalline domains because they lack sufficient regular order. Processing conditions such as extrusion cooking, heat-moisture treatment, or simply melting of the starch dispersion followed by cooling (as in temperature-cycling), might induce development of stable V-type assemblies, detectable by X-ray diffraction. The extent of V-structuring under these conditions is further dependent on starch type, and presence of organic guest molecules that can form inclusion complexes with the starch (Lopez-Rubio et al., 2007; Shamai et al., 2003).

Routinely, quantitative estimation of starch crystallinity is done from the peak intensities of X-ray diffractograms by fitting a baseline running through the minima of the individual peaks. A main inadequacy of this approach is that irregularly packed crystallites, small chain aggregates, minor extents of crystallisation, isolated single helices, and V-type structures, as already pointed out, are unable to diffract X-rays (Smits et al., 2003) and are therefore not accounted for. Furthermore, the intensity of peaks due to crystalline domains depends on the stability of atoms in their ideal position and deviations may lead to loss of reflection intensities. Under these circumstances, approaches that compensate for the limitations are a relevant consideration (Lopez-Rubio et al., 2008a; Mutungi et al., 2011a; Appendix 5). One alternative is provided in a method whereby a diffractogram of an amorphous sample is recorded, and adjusted using an appropriate correction factor to fit under the sample diffractogram (Mutungi et al., 2011a; Appendix 5). An illustration is given in Figure 11. In this method, the adjusted amorphous-sample diffractogram represents the amorphous component, and the crystalline component can be obtained from the total diffractogram by intensity subtraction. A further advantage of the method is that, by integrating the intensities of polymorph-specific peaks in the crystalline phase sub-diffractogram, proportions of the individual polymorphs can be approximated, thereby providing a quantitative estimation of long-range heterogeneity. A further drawback of X-diffraction in estimating crystallinity, also, relates to the effect of moisture. Whereas a certain amount of water is needed to maintain structural ordering of starch, sorption in the amorphous phase decreases the X-ray scattering relative to that of the real crystalline regions, and the plasticising effect of moisture further relieves inter-crystalline strains making intensities of diffraction peaks sharper or more pronounced, with a lower base line (Biliaderis, 2009).

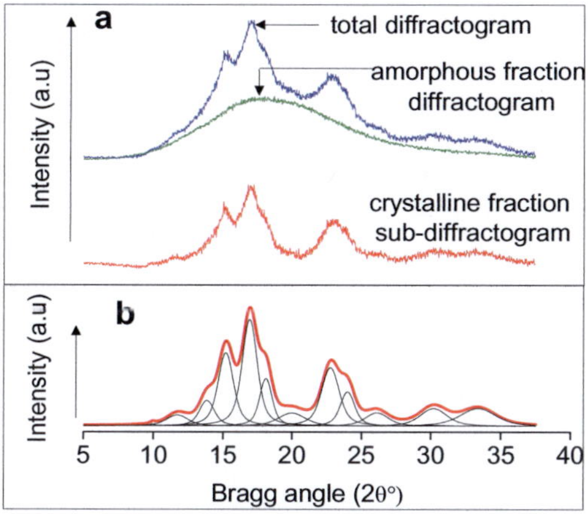

Figure 11: Estimation of starch crystallinity and proportions of the various polymorphs from respective Bragg peak intensities by decomposition of X-ray diffractogram into crystalline and amorphous components (a), and Gaussian function modelling of the crystalline fraction sub-diffractogram (b). The specific positions of Bragg peaks for A-, B- and V-polymorphs are cited in literature (Lopez-Rubio et al., 2008a; Mutungi et al., 2011a; Appendix 5).

In the majority of instances, crystallinity estimates of recrystallised starch preparations range between 25-70 % (Cairns et al., 1990; Gidley et al., 1995; Jacobasch et al., 2006; Pohu et al., 2004; Sievert et al., 1991), but higher crystallinity is also reported in preparations derived from *in vitro* synthesised linear α-polyglucans (Buleon et al., 2007). Generally, B-type preparations have relatively lower crystallinity values as compared to A-type preparations.

Characterisation of crystallinity is relevant to resistant starch because of the influence crystallinity has on susceptibility to digestive enzymes. However, the effects of starch digestion on crystallinity are also relevant because of the influence crystalline structure has on the physiological behaviour of resistant starch. Following digestion, crystallinity of recrystallised starch changes towards B-type due to chain disentangling effects of α-amylase, but in a manner related to the structural rigidity of the starting material (Mutungi et al., 2011a; Appendix 5). Controversy, however, exists regarding the effects of digestion on degree of crystallinity. Some observations suggest an increase (Cairns et al., 1990; Cairns et al., 1995; Lopez-Rubio et al., 2008b; Mutungi et al., 2011a; Appendix 5), and others a decrease (Fassler et al., 2006b) or no change at all (Faisant et al., 1995). These discrepancies are attributable to differences in the initial starch structure, the conditions of digestion, and method used to estimate crystallinity. Furthermore, starch that remains as enzyme-resistant fraction is the result of competition between the kinetics of digestion and the kinetics of molecular reordering of the starch as digestion progresses (Lopez-Rubio et al., 2008b).

2.7. Thermal characteristics

Application of differential scanning calorimetry to starch provides information relating to structure and interaction of the starch polymers. The heated starch gives rise to thermic transitions, whereby the corresponding enthalpy change is a quantitative measure of energy involved in chain association (exotherm) or dissociation (endotherm). Organisation of recrystallised starch is different in form and structure from that of native starch, and the conditions of recrystallisation, temperature in particular, influence association and interdependence of the amorphous and crystalline phases. Endothermic transitions, for example, become broader and the transition temperature lower, where recrystallisation temperatures are low because of poor chain ordering, although the rate of polymer re-association under such conditions is relatively rapid due to a high degree of heat withdrawal (Biliaderis, 1991). As a structural probe, thermal analysis by differential scanning calorimetry is sensitive to the molecular order regardless of the conformations of the polymer chains or their involvement in crystalline arrays, that is, enthalpy changes relate precisely to order-disorder transitions irrespective of the degree of crystallinity (Mutungi et al., 2011a; Appendix 5). Nevertheless, a direct relationship between crystallinity and thermal stability is expected (Figure 12) because of the involvement of the ordered polymers in crystalline domains.

Temperature and enthalpy of melting are both relevant parameters in defining structural stability of starch. Melting temperature reflects the density of packing of crystalline domains. Theoretically, closely packed A-type crystalline structures are more stable and therefore melt at a higher temperature than B-type structures. Occasionally, however, the temperature at which starch crystallites melt might not follow this rule because of the influence of factors such as crystallite size, length of the constituent chains, and the degree of crystallite perfection. Melting enthalpy, on the other hand, is associated with loss of double-helical order rather than loss of crystallinity, which infers that the forces responsible for the structural stability are largely at the double-helical level, and chain packing energy contributions are minimal. Even so, both crystalline and double-helical structure losses occur concurrently (Lopez-Rubio et al., 2008a) but loss of long-range order (crystallinity) might be expected to take place slightly earlier than loss of short-range order (double helices) during melting. Therefore distinction of the two processes, that is, crystallite melting and double helix dissociation, with regard to enthalpy, might be unrealistic.

In view of structural heterogeneity, a non-equilibrium nature of crystalline domains, and structural interdependence between amorphous and crystalline domains, broad and multiple melting transitions are typical of recrystallised starch (Figure 12). Multiple melting transitions may be a direct manifestation of distinctly stable ordered domains or composite thermal effects: disruption of ordered structures as well as their reordering during the heating process. In some instances, because of local structural heterogeneity, plasticisation gradients may also develop as heating progresses, resulting in heat flow-temperature profiles characterised by multiple transition peaks. Broad transition peaks are also consistent with a broad distribution of chain lengths in ordered or crystalline domains.

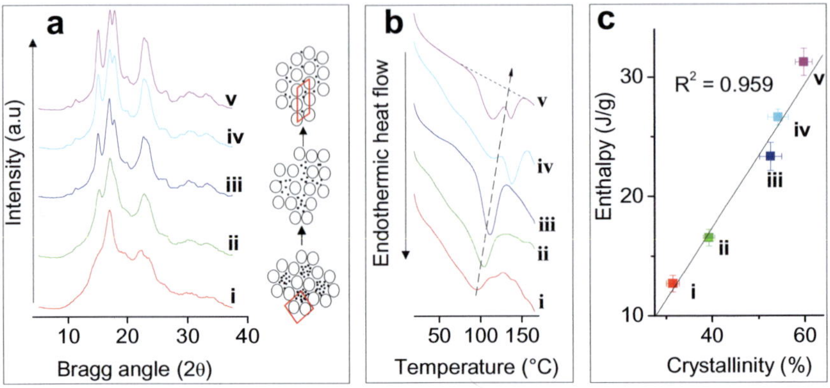

Figure 12: Crystalline transition and thermal properties of hydrothermally-treated freeze-dried debranched cassava starch powders. (a) X-ray diffraction pattern of debranched (i), debranched-annealed (ii), debranched-annealed and heat-moisture treated (iii), temperature-cycled (iv), and temperature-cycled and heat-moisture treated (v) forms. Model on the right depicts crystalline type transition due to progressive removal of water molecules (dots) and gradual sliding in of double helices. (b) Melting profiles of the starches; the slanting dashed arrow indicates increase in average peak melting temperature. (c) Correlation between crystallinity and melting enthalpy. Sample preparation conditions are: debranching, 10 % w/w starch suspension incubated with pullulanase (25 U/g starch) at 50 °C for 24 h; annealing, 20 % w/w starch gel held at 120 °C for 15 min, and then 60 °C for 24 h; heat-moisture treatment, starch containing 30 % w/w moisture held at 120 °C for 2 h; temperature-cycling, 20 % w/w starch gels subjected to three alternate cycles of 120 °C (2 h) and 60 °C (1 h) and then held at 60 °C for 15 h.

The relevance of thermal stability of recrystallised starch preparations is to be seen in the context of processing conditions of the foods into which fortification is intended. Processing and preparation of most foods also incorporates components such as sugars, salts and hydrocolloids. In conjunction with external factors such as moisture content, heating rate and shear or compressive forces, these ingredients are likely to impact on the melting behaviour of the starch. Understanding these interactions is important for better process control, improving the quality attributes of the fortified products, and ensuring availability of the functional starch. Sugars and polyhydric alcohols, in general, impede swelling and raise the temperature of melting, but have a restrictive effect on the re-ordering of starch polymers after melting. The effects arise from a combination of mechanisms: competition with the starch for water, anti-plasticisation, and *in situ* stabilisation of disordered chains into steady amorphous domains by building junctions with them. Interactions of starch chains with other polysaccharide hydrocolloids cause similar effects as sugars. The effects of salts, on the other hand, are more complex depending on the ionic character. A decline in melting temperature is observed in the presence of high concentrations of some salts, although the effect might be lower on A-crystalline type starch because of the dense structure. Some other salts cause thermal stabilisation by acting as dehydrating agents (Gidley, 1987).

Hydrothermal treatments provide a way of enhancing thermal stability by improving starch recrystallisation, and also introduce some desirable properties as far as thermal processing of the starch fortified foods is concerned. Alterations in solubility, swelling, and moisture interaction properties, among others, have relevant consequences (Mutungi et al., 2010; Appendix 4; Mutungi et al., 2011c; Appendix 6). Where the starch melting or pasting temperature is elevated, competition for moisture between starch and other macromolecules such as proteins undergoing denaturation during cooking processes is reduced. In baked goods, for instance, such a delay in starch pasting could mean a longer period of volume expansion, thus enhancement of desirable soft texture development. A reduction in swelling and solubility is also likely to increase the availability of water to other macromolecular components, sugars or salts in a composite food preparation. This quality is desirable where moisture is needed for specific technological or end product characteristics. Furthermore, under such circumstances, the stability of the functional starch is raised and so is the likelihood for its availability in the final product.

2.8. Structural conformation and molecular order

Conformations refer to three-dimensional arrangements of atoms in a molecule that result from rotation around a single bond, and define the relative energies of a molecule and therefore the relative reactivity and interactions with other molecules. For most starch-containing foods, the physical and functional properties can be traced to characteristics of the molecular species present and their interactions with each other, under the conditions (moisture, temperature, shear etc.) of processing and storage. Spectroscopic techniques such as infrared (IR) spectroscopy and nuclear magnetic resonance (NMR) provide means of exploring the molecular order and conformational organisation of starch based on the behaviour of atoms or atomic nuclei in their environments upon excitation with a source of external energy.

Fourier transform infrared absorption spectroscopy

IR techniques enable detection of changes in crystallinity and conformational or molecular order on the basis of frequencies and intensities of oscillation modes of chemical bonds (Table 2), arising from relative motions of atom nuclei in response to IR excitation. Widths and intensities of absorption bands reflect distribution and amount of bond energies. In the standard (absorption) Fourier transform-IR (FT-IR) spectroscopy, the wavelengths and intensity of absorption of mid-IR radiation, which is energetic enough to excite molecular vibrations to higher energy levels, is measured. Intensities of bands between wavenumbers of 1300 and 800 cm^{-1} are conformation-sensitive (van Soest et al., 1995) (Figure 13). In the disordered state, starch has a wider spread of conformations. With enhanced recrystallisation and molecular order, the range of conformations is reduced and a narrower distribution of bond energies is depicted in narrower absorption bands.

Table 2: Some band assignments for Infrared and Raman spectra of starch

Infrared	Infrared band assignment	Raman	Raman band assignment
537-711*	skeletal modes of pyranose ring	440-710	skeletal modes of pyranose ring
764	C-C stretching	763	C-C stretching
860	C(1)-H, CH_2 deformation	860	C(1)-H, CH_2 deformation
930	skeletal mode vibrations of α-1,4 glycosidic linkage, (C-O-C)	936	skeletal mode vibrations of α-1,4 glycosidic linkage, (C-O-C)
1067	C(1)-H bending	1087	C-O-H bending
1094	C-O-H bending	1122	C-O stretching, C-O-H bending
1163	C-O, C-C stretching	1200-1280	CH_2OH (side chain) related mode
1242	CH_2OH related mode	1139	C-O-H bending, CH_2 twisting
1344	C-O-H bending, CH_2 twisting	1382	CH_2 scissoring, C-H and C-O-H deformation
1415	CH_2 bending, C-O-O stretch	1460	CH_2 bending
1642	water in amorphous regions		
2800-3000	CH_2 deformation	2800-3000	C-H stretching
3000-3600	O-H stretching	3100-3600	O-H stretching

*wavenumber in cm^{-1} (Kizil et al., 2002; Mutungi et al., 2011b, Appendix 7)

To quantitatively evaluate FT-IR spectra, measurements are taken from band intensities using ratio measurements between selected peaks, typically, those exhibiting pronounced intensity changes. Of particular interest are the peaks at 1047 cm^{-1} and 1022 cm^{-1} characteristic of crystalline and amorphous starch, respectively; increased crystallisation or chain ordering is depicted by an increase in the ratio of intensities of these peaks (Mutungi et al., 2011a; Appendix 5).

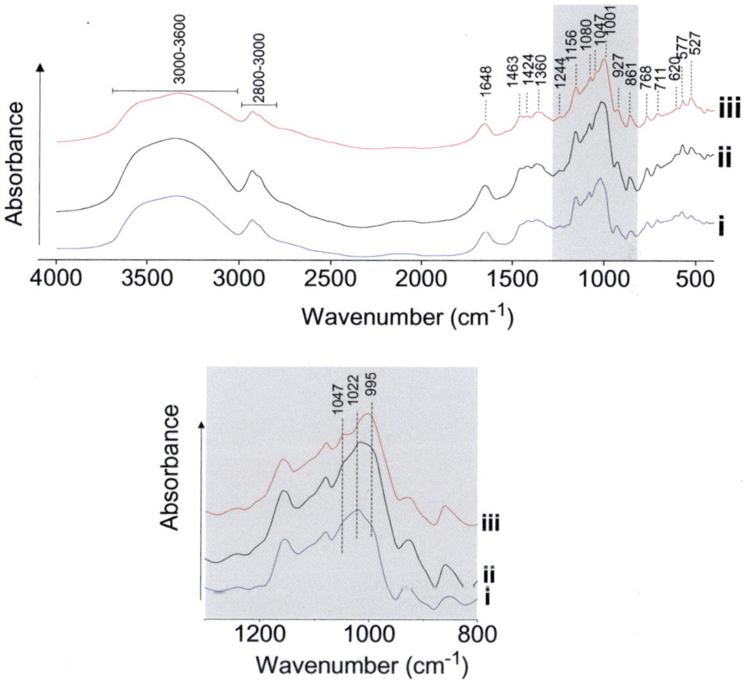

Figure 13: FT-IR spectra of freeze-dried amorphous (i), debranched-annealed (ii), and *in vitro* digested debranched-annealed cassava starch (iii), assigned with absorption bands. Variations in intensities and band narrowing at 1047, 1022 and 995 cm^{-1} depict change in molecular order. Starch subjected to digestion exhibits higher molecular order as compared to the undigested one (Mutungi et al., 2011a; Appendix 5). Sample preparation conditions are: amorphous, 10 % w/w native starch paste held at 120 °C for 15 min; debranching, 10 % w/w starch suspension incubated with pullulanase (25 U/g starch) at 50 °C for 24 h;debranched-annealed starch, 20 % w/w debranched starch gel held at 120 °C for 15 min, and then 60 °C for 24 h. Digestion: starch suspension incubated with a mixture of porcine pancreatic α-amylase and amyloglucosidase at 37 °C for 16 h.

Fourier Transform Raman spectroscopy

One shortcoming of infrared absorption spectroscopy is dominance of OH modes because water is a strong IR absorber. This means signals emanating from order or disorder associated modes are masked making interpretation difficult especially considering that water is integral to starch recrystallisation where it participates in building crystalline hydrate lattices. An alternative to FT-IR is FT-Raman (Figure 14). Unlike in FT-IR, water is a weak Raman scatterer and so vibrational modes are recorded as distinct peaks without strong combination bands. This enables a more accurate quantitative appreciation of molecular order (Mutungi et al., 2011b; Appendix 7). Fundamentally, Raman is similar to FT-IR spectroscopy, in that it also probes internal vibrations of molecules (Table 2). Raman spectroscopy however, relies on inelastic scattering of an incident photon, and characterises the bond motions in terms of energy required and the corresponding change in polarisability when the starch sample is focused with monochromatic light, usually from a laser in the near infrared range.

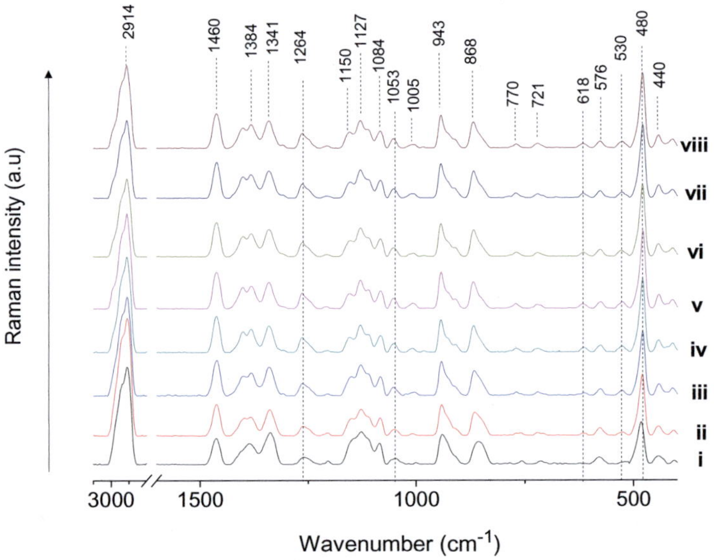

Figure 14: FT-Raman spectra of freeze-dried amorphous (i), debranched (ii), debranched-annealed (iii), debranched temperature-cycled (iv), and debranched heat-moisture treated (v) cassava starches, and the *in vitro* digested forms of debranched-annealed (vi), debranched temperature-cycled (vii), and debranched heat-moisture treated (viii) starches. Intensities of bands at 480, 530, 618, 1053 and 1264 cm^{-1} increase from i-viii when integrated with intensity of the band at 868 cm^{-1} as internal standard. Sample preparation conditions are: amorphous, 10 % w/w native starch paste held at 120 °C for 15 min; debranching, 10 % w/w starch suspension incubated with pullulanase (25 U/g) at 50 °C for 24 h; annealing, 20 % w/w starch gel held at 120 °C for 15 min, and then 60 °C for 24 h; heat-moisture treatment, 30 % w/w moisture containing starch held at 120 °C for 2 h; temperature-cycling, 20 % w/w starch gels subjected to three alternate cycles of 120 °C (2 h) and 60 °C (1 h), and then stored at 60 °C for 15 h. Digestion: starch incubated with a mixture of porcine pancreatic α-amylase and amyloglucosidase at 37 °C for 16 h.

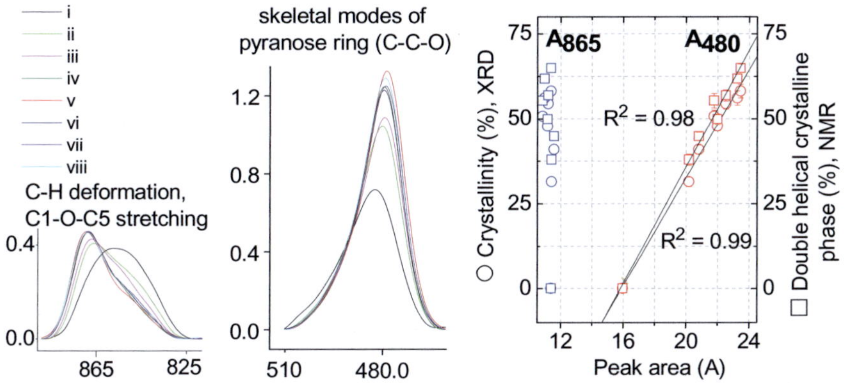

Figure 15: Correlation between peak area of the pyranose ring band vibrational mode at 480 cm^{-1}, and double helical crystalline phase (determined by ^{13}C CP/MAS NMR) and crystallinity (determined by XRD) of recrystallised and freeze-dried amorphous (i), debranched (ii), debranched-annealed (iii), debranched temperature-cycled (iv), debranched heat-moisture treated (v), *in vitro* digested debranched-annealed (vi), *in vitro* digested debranched temperature-cycled (vii), and *in vitro* digested debranched heat-moisture treated (viii) starches. Double helical content is constantly higher than crystallinity because some double helices do not form crystallites (Mutungi et al., 2011a; Appndix 5). Intensity of the band at 868 cm^{-1} does not change and is used as internal standard in integration of band intensity at 480 cm^{-1}. Sample preparation conditions are: amorphous, 10 % w/w native starch paste held at 120 °C for 15 min; debranching, 10 % w/w starch suspension incubated with pullulanase (25 U/g starch) at 50 °C for 24 h; annealing, 20 % w/w starch gel held at 120 °C for 15 min, and then 60 °C for 24 h; heat-moisture treatment, 30 % w/w moisture containing starch held at 120 °C for 2 h; temperature-cycling, 20 % w/w starch gels subjected to three alternate cycles of 120 °C (2 h) and 60 °C (1 h), and then stored at 60 °C for 15 h. Digestion: starch incubated with a mixture of porcine pancreatic α-amylase and amyloglucosidase at 37 °C for 16 h.

Generally, the starch spectrum is characterised by a strong band at 480 cm^{-1}, which is assigned to skeletal modes of the starch back bone (Kizil et al., 2002; Schuster et al., 2000). This strong skeletal mode band is not found in other polysaccharides, and the intensity is connected to the semi-crystalline nature of starch. The spectrum as well exhibits a typical band at approximately 865 cm^{-1} also called α-anomeric band of α-(1 → 4) linked glucopyranoses. This band associated with symmetric stretching of C(1)-O-C(5) bond of the glucopyranose ring (breathing mode), and the intensity should not change with the starch crystallinity or treatment, although the profile changes depending on the symmetry of ring vibration, for example, as influenced by helicalisation of the polymer chains. This band may therefore be used as internal standard to normalise experimental spectra for sample size. Intensities of a number of other Raman bands are also crystallinity sensitive (Mutungi et al., 2011b; Appendix 7). Raman spectroscopy seems to be a suitable technique for characterisation of the molecular and supramolecular structure of starch; intensity of the skeletal mode at 480 cm^{-1} is a good index for molecular order and may be used to estimate the crystallinity of starch preparations (Figure 15) particularly if samples have a common or comparable polymer chain structure (Mutungi et al., 2011b; Appendix 7).

Nuclear magnetic resonance spectroscopy

Nuclear magnetic resonance (NMR) techniques for starch structure elucidation centre on evaluation of either the rate of free induction decay (time-domain NMR), or the frequencies at which free induction decay signals resonate (frequency-domain NMR) when atomic nuclei that have a nuclear spin (magnetic moment) are excited with a radio frequency pulse in a strong magnetic field. ^{1}H and ^{13}C NMR are particularly common. On the one hand, time-domain ^{1}H NMR elucidates starch physical structures from the micro-dynamic aspect of the starch molecules (Ablett, 1992). Obviously, the initial intensity of the signal is directly proportional to the number of nuclei present, but the signal decay rate once the electromagnetic pulse is withdrawn is related to molecular mobility. Differences in molecular mobility are reflected in the spin-spin relaxation time (T_2), because this parameter is several orders of magnitude lower in components that have restricted mobility as compared to the relatively mobile ones. This way, amorphous (more mobile) and crystalline or ordered (less mobile) starch segments are distinguished.

^{1}H NMR may also be applied in the frequency domain. The basis is that the exact resonance frequency of a nucleus depends on the physical and chemical composition of the atomic space surrounding the nucleus, which includes the electron clouds of an atom, those of surrounding atoms, and the atomic bonding. This influence is called chemical shift of the resonance frequency and the relative difference in resonant frequency is referenced against a standard signal, usually that of tetramethylsilane (has “zero” chemical shift), and expressed in ppm since the numerator is in Hertz and the denominator is in Megahertz. However, although ^{1}H NMR is highly sensitive and ^{1}H is an abundant spin species (natural abundance is approximately 99.98 %), signals are often crowded within a narrow chemical shift range, and this lessens the suitability for delineation of structural features because of signal-overlap (Ablett, 1992).

^{13}C NMR, on the other hand, has a much weaker signal because ^{13}C is a dilute spin species (natural abundance is approximately 1.1 %) and is also less sensitive as compared to ^{1}H (sensitivity relative to ^{1}H is 0.0159). Nevertheless, cross-polarisation (CP) and magic-angle spinning (MAS) applications enhance sensitivity and signal resolution, respectively, enabling starch conformations to be deduced on the basis of their motional effects in solid domains (solid state ^{13}C CP/MAS NMR). An additional advantage of ^{13}C NMR over ^{1}H NMR is that chemical shifts are spread over a broader range, and independent resonance peaks are obtained for each carbon atom without multiplicity complications arising from ^{13}C-^{13}C coupling. These characteristics make ^{13}C CP/MAS NMR more desirable in marking out structural conformations (Ablett, 1992; Bertocchi & Paci, 2008). Chemical shifts of individual carbon atoms depend on the screening of external magnetic field by local electron densities. In particular, signal of C1 atom responds not only to bonding structure but also to geometrical orientations of bonds due to conformation related twisting of chains about α-(1→ 4) glycosidic linkage (Figure 16), hence chain conformations can be evaluated from multiplicity patterns in this particular signal (Gidley & Bociek, 1988; Morgan et al., 1995; Paris et al., 1999).

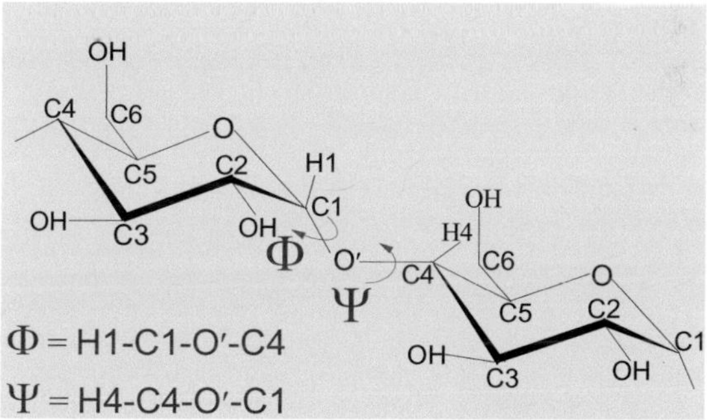

Figure 16: α-(1 → 4) glycosidic bond dihedral (torsion) angels, Ψ and Φ that characterise structural conformations of starch chains (Paris et al., 2001)

Starch is essentially amorphous subsequent to gelatinisation, and reorganises into diverse low energy morphologies during debranching. These structures become further stabilised when a gel of the debranched starch or a wetted powdered sample is subjected to hydrothermal treatments involving defined moisture and temperature conditions (Mutungi et al., 2009b; Appendix 3). From a qualitative point of view, the structural modifications arising from these treatments are reflected in the C1 and C4 signal profiles (Figure 17). Furthermore packing patterns of chain conformations in crystalline domains (polymorphism) are distinguished from peak multiplicities of C1 signal. A triplet represents A-type crystallinity, doublet B-type crystallinity, and a single resonance peak at 103 ppm is indicator for V-type crystallinity (Paris et al., 1999; Snape et al., 1998; Tan et al., 2007). Unlike ^{1}H NMR, however, the relative strengths of ^{13}C NMR signals are not related to the number of atoms generating each one of them but to cross-polarisation dynamics.

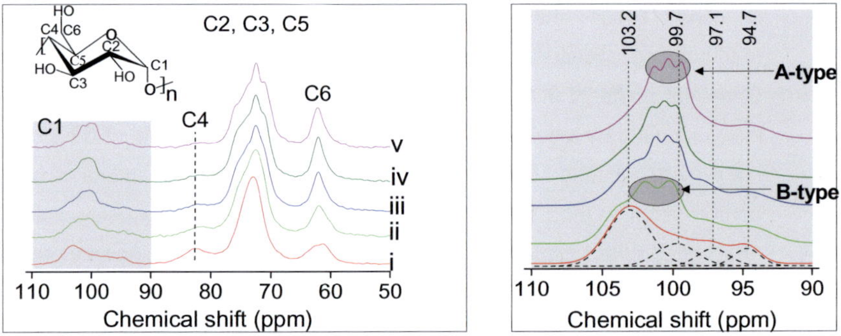

Figure 17: ^{13}C CP/MAS NMR spectra of freeze-dried amorphous (i), debranched (ii), debranched-annealed (iii), debranched temperature-cycled (iv), and debranched heat-moisture treated (v) cassava starch. On the right, C1 signal profiles are zoomed in to portray local short-range heterogeneity and crystalline type. Peak multiplicities are not fully resolved in the recrystallised starches probably because crystalline domains are not perfectly packed into pure polymorphic forms (Mutungi et al., 2011a; Appendix 5). Sample preparation conditions are: amorphous, 10 % w/w native starch paste held at 120 °C for 15 min; debranching, 10 % w/w starch suspension incubated with pullulanase (25 U/g starch) at 50 °C for 24 h; annealing, 20 % w/w starch gel held at 120 °C for 15 min, and then 60 °C for 24 h; temperature-cycling, 20 % w/w starch gels subjected to three alternate cycles of 120 °C (2 h) and 60 °C (1 h), and then stored at 60 °C for 15 h; heat-moisture treatment, 30 % w/w moisture containing starch held at 120 °C for 2 h.

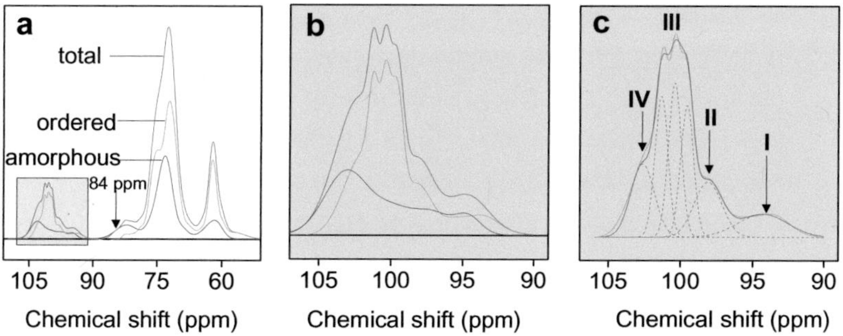

Figure 18: Estimation of short-range heterogeneity from ^{13}C CP/MAS NMR spectra: (a) decomposition of the total spectrum into ordered and disordered sub-spectra by subtraction of an adjusted amorphous-sample spectrum such that the intensity of the difference spectrum is zero at 84 ppm; (b) profiles of C1 signal in the amorphous, ordered and total spectra; and (c) deconvolution of the ordered sub-spectrum C1 signal into peaks corresponding to constrained conformations (I), non-crystalline double helices (II), crystalline double helices (III), and single-helical or V-type conformations (IV) (Mutungi et al., 2011a; Appendix 5).

Splitting or multiplicity evolution in the C1 atom signal is governed by packing constraints, meaning that multiplet spectral signatures at C1 atom signal are not attributed to a general molecular order (helicity) but to crystallinity it self (Paris et al., 1999). Purely well crystallised double helices yield narrow distributions of the conformational angles associated to their respective chemical shift, while non-crystallographically constrained chain ends, as well as defective chains at unpacked boundaries or internal defects should be sufficiently distorted on an average so as not to show up at similar chemical shifts. As a result of crystallographic packing constraints, two sets of established symmetry rules are normally applied, which correspond to A- and B-type space groups (Imberty et al., 1991; Paris et al., 1999). In A-type structure, the crystallographic two fold axis (arising from maltotriose as repeat unit in the double helices) produces three equivalence classes among the six monomers that constitute a single helical turn. The three equivalence classes correspond to three different α-(1 → 4) linkages i.e. three Φ and Ψ torsion angle combinations, hence a triplet is observed on the C1 signal. On the contrary, in B-type structure, a three fold helical packing symmetry (arising from maltose as repeat unit) produces two equivalence classes corresponding to two different α-(1 → 4) linkages, hence a doublet is observed.

Estimations of the relative proportions of various structural conformations, and quantification of local short-range heterogeneity, may thus be achieved by application of curve-fitting algorithms or deconvolution techniques to derive intensities of distinct peaks (Figure 18) (Mutungi et al., 2011a; Appendix 5). By comparing peak areas of C1 signal in the ordered sub-spectrum relative to those of the total spectrum, the relative proportion of amorphous, double-helical, single-helical and constrained conformations in the starch sample are obtained. The deduction of ordered as well as non-ordered double- and single-helical conformations makes it further possible to estimate proportions of structures that are in or out of crystalline registry (Mutungi et al., 2011a; Appendix 5, Mututngi et al., 2011b; Appendix 7).

A concern that arises regarding this quantitative approach, however, is the appropriateness when comparing domain heterogeneity of differently prepared samples, as differences in the local environments of similar atoms might cause non-uniform local polarisation effects, and therefore signal inconsistencies (Tan et al., 2007). In principle, CP/MAS enables acquisition of signals from the carbon sites experiencing incompletely averaged H-C dipolar interactions. Where local static dipolar fields are weak due to increased segmental mobility or lack of nearby protons, the build-up of polarisation will be slow, and thus, the corresponding ^{13}C signal intensity will also be low in the CP/MAS spectrum. Generally C1-C5 atoms are considered to be relatively rigid sites, and the intensities are expected to be fairly invariable when short durations of cross polarisation are used as compared to C6 atom whose signal intensity is more sensitive because of the local mobility associated with this exocyclic group (mobility results in a low rate of cross polarisation due to incomplete averaging of the heteronuclear dipolar couplings) (Tan et al., 2007). A cross-polarisation contact time that ensures relative constancy in local magnetisation effects is usually chosen (Paris et al., 1999; Tan et al., 2007). A test that may be applied to confirm internal consistency of spectra involves checking that the ratios of the combined intensities due to C2-C5 atoms and intensity due to C1 atom is the same across the spectra. If this criterion is met, the amorphous and crystalline regions behave similarly with respect to their solid state nature, and spectra of different samples may be analysed quantitatively, and C1 signal data, in particular, compared (Mutungi et al., 2011a; Appendix 5;

Mutungi et al., 2011b; Appendix 7; Paris et al., 1999; Tan et al., 2007).

2.9. Starch digestibility

Reasons for resistance of starch to enzyme hydrolysis

Digestion of starch is carried out by α-amylase and glucoamylase. The action of α-amylase [EC 3.2.1.1] results in maltose, longer chain oligosaccharides, and some glucose. Glucose is absorbed directly through the intestinal mucosa, whereas maltose and the oligosaccharides are hydrolysed further by membrane-bound glucosidases that include glucoamylase (amyloglucosidase). Glucoamylase [EC 3.2.1.3] hydrolyses both α-D-(1 → 4) and α-D-(1 → 6) glycosidic bonds. Starch digestibility is, however, predominantly dependent on the action of α-amylase, which hydrolyses α-D-(1 → 4) glycosidic bonds by an endo-mechanism, retaining the α-configuration at the reducing end anomeric carbon in the products (van der Maarel et al., 2002). Multiple-attack mechanism is a significant feature of α-amylolysis. In this mechanism, a glucan chain that binds onto the catalytic cleft undergoes several bond splits before the final fragment completely dissociates from the site. The binding site also comprises several sub-sites (Figure 19).

Thus one important defining aspect of α-amylase activity is the number of individual bonds split for every productive encounter that a starch chain makes with the active-site. Other enzyme factors that influence the nature and rate of starch α-amylolysis include the number of binding sub-sites in the catalytic cleft, their affinity for glycosyl units, and the position of the catalytic groups that perform the actual bond hydrolysis. Alpha-amylases from different sources differ in the number of glucosyl-unit binding sub-sites (Figure 19), and shorter chain products are formed if an enzyme has fewer binding sub-sites. Similarly, the frequency of bond splitting (turnover) is higher for larger maltodextrins compared to maltotrioses and maltotetraoses, because the smaller molecules reside longer at the binding sites, which also explains, in part, why rate of α-amylolysis diminishes as hydrolysis advances (Robyt, 2009). These factors are particularly relevant to the choice of enzyme for *in vitro* starch digestibility investigations.

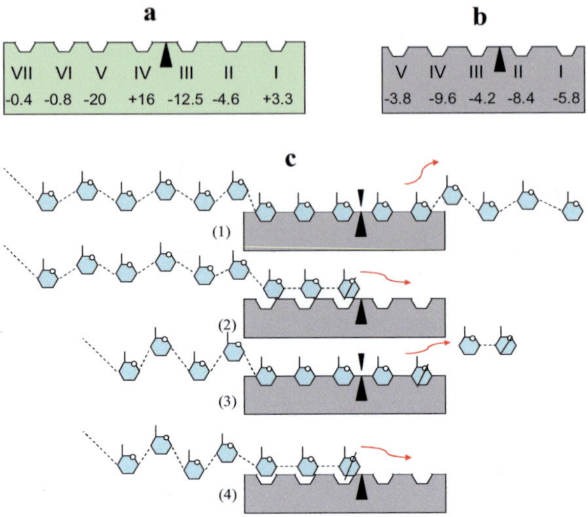

Figure 19: Simplified illustration of glycosyl unit binding sub-sites at the catalytic cleft of two different α-amylases: α-amylase from *Aspergillus oryzae* with seven binding sub-sites (a) and α-amylase from porcine pancreas with five binding sub-sites (b). Sub-sites are numbered with Roman numerals beginning with the sub-site that would bind the reducing-end glycosyl unit. Arabic numbers indicate binding energies in kJ/mol; increasing negative values depict stronger binding. The symbol (▲) represents location of the catalytic groups in relation to the glucose binding sites. Part (c) illustrates multiple-attack of α-glucan chain bound to the active site of porcine pancreatic α-amylase. In step (1), the chain is cleaved and the right-hand chain diffuses away from the active site (2). The chain that remains at the active site is then shuttled to fill the free sub-sites (3), is cleaved to give maltose or maltotriose in the case of *A. oryzae* α-amylase, and the process repeats (4) or the chain dissociates from the active site (Robyt, 2009).

Structure of starch substrates also defines digestion kinetics in a fundamental way. In the case of native starch granules, attack by α-amylase is, among other factors, subject to granular size, surface properties (surface pores, surface proteins and surface lipids), crystallinity, branching structure, ratio of amylose to amylopectin, and the amount of amylase-lipid complexes (Singh et al., 2010). Amylose-rich starches and B-type starches, in general, show higher resistance to digestion because they have higher amounts of longer double-helix forming amylopectin chains, and therefore higher double-helical content. Location of α-D-(1 → 6) branch points in the crystalline regions weakens the crystalline structure, enhancing penetration of the enzyme into the granule interior (although kinetically, bond hydrolysis slows down as the branching points are approached). High lipid contents result in high V-type structures that sterically hinder enzyme binding.

Surface properties that relate to hydrophilic character affect the diffusion of α-amylase from the liquid or bulk-phase to the solid-liquid interface, and subsequently, its adsorption onto the (solid) substrate.

Resistance of recrystallised starch to digestion is due to ordering of starch molecules into double helices that eventually rearrange into dense supramolecular assemblies stabilised by hydrogen bonding (Cairns et al., 1990; Pohu et al., 2004). At the long-range, a crystalline structure hinders diffusivity and adsorption of the enzyme onto the solid substrate (Leloup et al., 1991), meaning that enzyme and substrate cannot approach each other over sufficient surface (Mutungi et al., 2011c; Appendix 6). This hindrance may depend on properties of the crystallites such as length of constituent chains (stability of helical structures), packing density of the chains, chain orientation (conformation), proximity of amorphous regions in relation to the crystalline ones, and accessibility of chain extremities.

At the short-range, double helical structures resist digestion because they cannot be accommodated in the active site (Casset et al., 1995). Helicalisation also means that only three glucose unit segments are linearly aligned which is insufficient for complete sub-site binding (human salivary and pancreatic α-amylases consist of 5-7 and 5-6 sub-sites, respectively) and therefore formation of a productive enzyme-substrate complex is not attained. Attainment of the half-chair conformation by glycopyranosyl units, which is important for bond cleavage, is also constrained in highly ordered domains. These arguments are supported by the high double helical content and molecular order determined in polymeric residues of *in vitro* starch digestion (Mutungi et al., 2011a; Appendix 5). Double helices are stabilised by hydrogen bonding, yet for optimal α-amylase action, a series of glycosyl units need to fill all binding sub-sites of the catalytic cleft (Figure 19). Starch chains that are intimately associated into double helices are unlikely to bind effectively on the catalytic cleft unless they have been disentangled. However, where double helices or crystalline regions contain imperfections, the chains in these regions may bind the active site, and attack on these zones can be expected to cause local destabilisation of the double helical and crystalline order in favour of continued amylolysis. Nevertheless, depending on proximity of the amorphous regions in relation to the crystalline ones, hydrolysis is likely to be relatively slow because after one bond is hydrolysed, the enzyme has to reposition the glucan chain so that all sub-sites are newly occupied, alternatively the enzyme molecule can associate with another section of the starch chain.

Binding sub-sites also have relevant implications on the size of polymeric residues of starch digestion in the sense that bond cleavage will reach a limit several glycosyl units before the point where the ordered regions begin. If an enzyme has a large number of binding sub-sites, longer stubs remain and vice versa. This is particularly a relevant consideration in comparing *in vitro* estimates of resistant starch chain length distributions. On the average, the most highly enzyme resistant polymer chains in recrystallised debranched cassava starch are those comprising 18-48 glucose units (Mutungi et al., 2011a; Appendix 5). A wider distribution of polymer chains in the digested starch is, however, to be expected because of reasons related to the starch structure and enzyme factors discussed above.

Modelling starch digestibility

Kinetic models and digestibility indices are often used to describe and predict starch digestion (Dona et al., 2010; Singh et al., 2010). Kinetic models provide a means of predicting how the enzyme binds on the substrate, turns the substrate into products and the rate and efficiency of the processes. The choice of kinetic models is informed by the kind of information desired. In view of the classical Michaelis-Menten kinetics, velocity of α-amylase action on starch is linear with time during the early stages, and can be related to the initial starch concentration [S] as $v = v_{max} \cdot [S]/K_m + [S]$, where v_{max} is maximum reaction velocity and K_m is Michaelis constant. $v_{max} = k_{cat} \cdot [E]_0$ where K_{cat} is catalytic rate constant or turnover number and $[E]_0$ is total (initial) enzyme concentration. For a given starch substrate, K_m is the concentration of readily digestible starch $[S_d]$ that will support an initial reaction rate equivalent to $v_{max}/2$ (Tahir et al., 2011). If [S] is not limiting and the ratio $[S]/[S_d]$ is large, K_m can be measured and starch treatments that increase or lower $[S_d]$ will be reflected in the K_m value, as K_m is indicator of the affinity and extent of enzyme binding to the substrate (Tahir et al., 2011). A low K_m will indicate a higher affinity for the substrate, a more stable enzyme binding and therefore a more effective catalysis, whereas a higher K_m is indicator for less effective catalysis due to loss of structural complementarities following the treatments. Measurements of K_m might therefore provide a direct indication of the starch fraction that can be readily digested, and so be applicable in examining starch structure-digestibility relationships.

K_{cat} gives the number of product molecules formed per molecule of enzyme per unit time. In the case of α-amylase action on starch, K_{cat} is expected to attain the maximum possible value when access to glycosidic linkages is not hindered at all by structural constraints, water molecules for the hydrolytic reaction are not limiting, and the reaction products can freely exit the catalytic site. These conditions are only likely when the enzyme acts upon completely-gelatinised amorphous starch. The ratio K_{cat}/K_m, also known as specificity constant or catalytic efficiency, is substrate specific and is certainly a means for comparing reactivity of α-amylase towards differently treated starches as it expresses the affinity and catalytic ability of the enzyme towards the substrate (Tahir et al., 2011); specificity constant is low when enzyme and substrate cannot freely approach each other.

Reliable estimates of Michaelis-Menten parameters are, however, obtainable only when rate measurements are made within the early stages of a reaction, where product formation or substrate disappearance is linear with time. Hydrolysis of starch often exhibits complications that arise from structural alterations and product inhibition as well, factors that might annul the linearity assumption. It also becomes difficult to attribute the kinetic parameters (K_m for example) to the original structural state of the starch. Furthermore, *in-vitro* starch digestibility studies are performed with high enzyme concentrations over prolonged incubation times in order to simulate *in vivo* digestibility as closely as possible.

Digestibility indices such glycemic response (GR), glycemic index (GI), glycemic load (GL) and resistant starch are more commonly used because they reflect availability of the starch *in vivo*. GR is a measure of the increase in blood glucose concentration with time, after the ingestion of a carbohydrate meal. GI is the total glycemic response in blood in the first 2 h after ingestion of a fixed amount of carbohydrate relative to that of a standard food, usually white bread or glucose, whereas GL is GI adjusted for percent carbohydrate content of a food, and is useful in clustering similar foods. Because conducting *in vivo* experiments to estimate these parameters is expensive, a first order exponential kinetic model proposed by Göni and co-workers (Goni et al., 1997) is commonly used to analyse *in vitro* digestibility data. From the model, $C = C_{\infty}(1 - e^{-kt})$, where C is concentration of glucose at time *t*, C_{∞} is glucose concentration at "end" of reaction and *k* is a first order rate constant, different starch substrates give different values for *k* that reflect their relative susceptibilities to digestion. C_{∞} is a measure of resistant starch, that is, the amount of starch remaining when further product formation is negligible under standard digestion conditions. Moreover, by relating the area under the model digestibility curve of a sample to the area under the model digestibility curve of a reference food, such as white bread or amorphous starch, the relative digestibility or hydrolysis index is obtained, which correlates well with *in vivo* GI data (Goni et al., 1997). A detailed overview of methods for resistant starch determination was provided by Champ and co-workers (Champ et al., 2003). Although the method by Göni *et al.,* (Goni et al., 1997) is widely acclaimed, numerous other methods have also been used in the original or modified form, and a most befitting approach in resistant starch measurements for now, seems to be comprehensive description of the conditions applied so as to allow comparison with other studies.

3. Conclusion and outlook

Dietary components escaping digestion in the human gut contribute to health, well-being and illness prevention, especially if these components are fermentable by gut microbiota. Certain fractions of dietary starch are able to avoid action of digestive enzymes, and considerable amounts are expected to reach the lower gut. Because many modern convenience foods are rich in highly digestible components, consumption of functional fibre in general is still low among populations where tendencies to consume highly processed foods are high. The per capita daily intake of resistant starch is estimated at 5 g, 5-7 g, 3-6 g and 10-18 g in the USA, Australia, Europe, and India/ China, respectively, whereas a daily intake of at least 20 g is recommended in order to attain the benefits ascribed to it. Consequently development of resistant starch preparations that can be readily incorporated in frequently consumed foods as functional ingredients or made available as supplementation nutraceutical products is valid. Recrystallised starch offers a practical possibility for availing a functional starch ingredient, whose techno-functional properties can further be optimised for suitability in a number of products.

An objective of this study was to investigate methods of increasing resistant starch formation by way of starch recrystallisation. The strategy of applying debranching enzymes to convert aqueous starch suspensions into linear polymers enables the starch to re-associate quickly into partial-order domains. Structural stability is improved variously by processing the debranched starch under selected physical conditions involving heat and moisture such as annealing, temperature-cycling and heat-moisture treatment. The heat and moisture regimes in these treatments are also encountered in processing of various foods. Constraints imposed by localised crystallisation or structural order slow down susceptibility of the starch to digestive enzymes. Heat-moisture treatment offers the optimal approach, and perfects the build-up of rigid crystalline domains that, among other factors, impede diffusivity of digestive enzymes into the solid substrate. Consequently, a product containing > 80 % resistant starch is obtained by heat-moisture treatment of the debranched starch. The behaviour of this preparation in processed products and the physiological functionality, particularly the interaction with microbial populations associated with the human gut, need to be investigated.

For most starch-containing foods, the physical and functional properties can be traced to the characteristics of molecular species present, their interactions with each other, and the modifications that may be caused by conditions such as moisture, temperature, and shear during processing and storage. A further objective of this study was to investigate the molecular and supra structure of the recrystallised starch because chain organisation and conformational order also govern the interactions between the starch and other food constituents. A combination of methods was used to achieve a comprehensive structure analysis. Different techniques do not detect the same properties, or at best, are not sensitive to the same structural elements. Thermal analysis is sensitive to chain ordering as manifested by an endothermic peak, and X-ray diffraction measures long-range crystalline order. FT-IR and Raman spectroscopy probe conformation- and order-dependent molecular vibrations, whereas NMR detects segmental motions, conformational-dependent chemical shifts and short-range order.

On the basis of findings of this study, recrystallised debranched cassava starch comprises structures in various metastable states, that is, both ordered and amorphous phases. The relative abundances are dependent on the method chosen to improve the recrystallisation that already starts during debranching. Ordered conformations are compartmentalised into crystalline double helices, non-crystalline double helices, constrained conformations and single helices, whose relative contents also vary depending on the recrystallisation method. Temperature-cycling and heat-moisture treatment favour packing of double helices into A-type crystalline form, whereas the milder annealing process retains a higher proportion of the B-type morphology. Only small amounts of V-type crystalline structure are discernible in the recrystallised starch. Crystallinity certainly plays a role in the resistance of the starch to digestion but amorphous conformations that are caged within imperfect crystals, non-crystalline double-helical structures, single helices, and constrained conformations are also likely to restrict or favour digestion in a manner dependent on their proximity to the crystalline regions.

Physical and molecular structure of resistant starch also has influence on the interaction with gut microbiota. *In vitro* digestion enhances crystallinity and molecular order of recrystallised starch. A-type crystalline packing decreases whereas the B- and V-type morphologies increase, and the disordered phase diminishes. The digested starches are however structurally less rigid, and the stoichiometry of the resultant conformations is a function of the crystalline rigidity of the initial material. The A-type preparation by heat-moisture treatment of debranched starch is apparently more resistant to structural alteration because of lower enzyme attack. This is advantageous in that significant amounts of the starch can be expected to reach the lowest parts of the gut where maintenance of colorectal integrity by butyrate producers is more relevant. With regard to chain distribution, helical chains made up of 18 - 48 glucose units are most resistant to digestion, but a broad distribution of chain lengths is likely to be encountered in the polymeric residue of starch digestion depending on the extent to which α-amylase can access and hydrolyse the glucan chains.

Finally, quantitative characterisation of recrystallised starch structure was explored using a number of methods. These approaches can be applied to follow quantitatively the structure and stability of other starch-based substrates upon physical, chemical or enzymatic conversions as well. Structural and phase transitions are also common in industrial practice, which raises perspectives of widening the approaches for monitoring processes or products relating not only to starch but to other materials especially of polymeric origin, where state transitions are relevant to product quality, stability, performance or functionality.

4. Bibliography

Ablett, S. (1992). Overview of NMR applications in food science. *Trends in Food Science & Technology*, **3**, 246-250.

Asp, N. G. (1992). Resistant Starch - Proceedings from the 2nd plenary meeting of Euresta - European Flair Concerted Action 11 on Physiological implications of the consumption of resistant starch in man. *European Journal of Clinical Nutrition*, **46**, S1-S1.

Belenguer, A., Duncan, S. H., Calder, A. G., Holtrop, G., Louis, P., Lobley, G. E., & Flint, H. J. (2006). Two routes of metabolic cross-feeding between *Bifidobacterium adolescentis* and butyrate-producing anaerobes from the human gut. *Applied and Environmental Microbiology*, **72**, 3593-3599.

Bertocchi, F., & Paci, M. (2008). Applications of high-resolution solid-state NMR spectroscopy in food science. *Journal of Agricultural and Food Chemistry*, **56**, 9317-9327.

Bertoldo, C., & Antranikian, G. (2002). Starch-hydrolyzing enzymes from thermophilic archaea and bacteria. *Current Opinion in Chemical Biology*, **6**, 151-160.

Biliaderis, C. G. (1991). The Structure and interactions of starch with food constituents. *Canadian Journal of Physiology and Pharmacology*, **69**, 60-78.

Biliaderis, C. G. (2009). Structural transitions and related physical properties of starch. In: J. Bemiller, & R. Whistler, *Starch : Chemistry and Technology* (pp. 293-372). Amsterdam: Elsevier.

Bird, A. R., Brown, I. L., & Topping, D. L. (2000). Starches, resistant starches, the gut microflora and human health. *Current Issues in Intestinal Microbiology*, **1**, 25-37.

Bouhnik, Y., Raskine, L., Simoneau, G., Vicaut, E., Neut, C., Flourie, B., Brouns, F., & Bornet, F. R. (2004). The capacity of nondigestible carbohydrates to stimulate fecal bifidobacteria in healthy humans: a double-blind, randomized, placebo-controlled, parallel-group, dose-response relation study. *American Journal of Clinical Nutrition*, **80**, 1658-1664.

Brouns, F., Arrigoni, E., Langkilde, A. M., Verkooijen, I., Fassler, C., Andersson, H., Kettlitz, B., van Nieuwenhoven, M., Philipsson, H., & Amado, R. (2007). Physiological and metabolic properties of a digestion-resistant maltodextrin, classified as type 3 retrograded resistant starch. *Journal of Agricultural and Food Chemistry*, **55**, 1574-1581.

Brown, I., Warhurst, M., Arcot, J., Playne, M., Illman, R. J., & Topping, D. L. (1997). Fecal numbers of bifidobacteria are higher in pigs fed *Bifidobacterium longum* with a high amylose cornstarch than with a low amylose cornstarch. *Journal of Nutrition*, **127**, 1822-1827.

Brown, I. L., Wang, X., Topping, D. L., Playne, M. J., & Conway, P. L. (1998). High amylose maize starch as a versatile prebiotic for use with probiotic bacteria. *Food Australia*, **50**, 603-610.

Buleon, A., Colonna, P., Planchot, V., & Ball, S. (1998). Starch granules: structure and biosynthesis. *International Journal of Biological Macromolecules*, **23**, 85-112.

Buleon, A., Veronese, G., & Putaux, J. L. (2007). Self-association and crystallization of amylose. *Australian Journal of Chemistry*, **60**, 706-718.

Burkitt, D. P. (1973). Epidemiology of large bowel disease: the role of fibre. *Proceedings of the Nutrition Society*, **32**, 145-149.

Cairns, P., Leloup, V., Miles, M. J., Ring, S. G., & Morris, V. J. (1990). Resistant starch - an X-ray diffraction study into the effect of enzymatic-hydrolysis on amylose gels *in vitro*. *Journal of Cereal Science*, **12**, 203-206.

Cairns, P., Sun, L., Morris, V. J., & Ring, S. G. (1995). Physicochemical studies using amylose as an *in-vitro* model for resistant starch. *Journal of Cereal Science*, **21**, 37-47.

Casset, F., Imberty, A., Haser, R., Payan, F., & Perez, S. (1995). Molecular modeling of the interaction between the catalytic site of pig pancreatic α-amylase and amylose fragments. *European Journal of Biochemistry*, **232**, 284-293.

Champ, M., Langkilde, A. M., Brouns, F., Kettlitz, B., & Collet, Y. L. (2003). Advances in dietary fibre characterisation. 1. Definition of dietary fibre, physiological relevance, health benefits and analytical aspects. *Nutrition Research Reviews*, **16**, 71-82.

Charles, A. L., Sriroth, K., & Huang, T. C. (2005). Proximate composition, mineral contents, hydrogen cyanide and phytic acid of 5 cassava genotypes. *Food Chemistry*, **92**, 615-620.

Charoenkul, N., Uttapap, D., Pathipanawat, W., & Takeda, Y. (2006). Molecular structure of starches from cassava varieties having different cooked root textures. *Starch-Starke*, **58**, 443-452.

Cheetham, N. W. H., & Tao, L. P. (1998). Variation in crystalline type with amylose content in maize starch granules: an X-ray powder diffraction study. *Carbohydrate Polymers*, **36**, 277-284.

Crittenden, R. G. (1999). Prebiotics. In: G. W. Tannock, *Probiotics: A Critical Review* (pp. 141-156). Wymondham: Horizon Scientific Press.

Crittenden, R. G., Morris, L. F., Harvey, M. L., Tran, L. T., Mitchell, H. L., & Playne, M. J. (2001). Selection of a Bifidobacterium strain to complement resistant starch in a synbiotic yoghurt. *Journal of Applied Microbiology*, **90**, 268-278.

Cummings, J. H., Beatty, E. R., Kingman, S. M., Bingham, S. A., & Englyst, H. N. (1996). Digestion and physiological properties of resistant starch in the human large bowel. *British Journal of Nutrition*, **75**, 733-747.

Dona, A. C., Pages, G., Gilbert, R. G., & Kuchel, P. W. (2010). Digestion of starch: *In vivo* and *in vitro* kinetic models used to characterise oligosaccharide or glucose release. *Carbohydrate Polymers*, **80**, 599-617.

Dongowski, G., Jacobasch, G., & Schmiedl, D. (2005). Structural stability and prebiotic properties of resistant starch type 3 increase bile acid turnover and lower secondary bile acid formation. *Journal of Agricultural and Food Chemistry*, **53**, 9257-9267.

Duncan, S. H., Holtrop, G., Lobley, G. E., Calder, A. G., Stewart, C. S., & Flint, H. J. (2004a). Contribution of acetate to butyrate formation by human faecal bacteria. *British Journal of Nutrition*, **91**, 915-923.

Duncan, S. H., Louis, P., & Flint, H. J. (2004b). Lactate-utilizing bacteria, isolated from human feces, that produce butyrate as a major fermentation product. *Applied and Environmental Microbiology*, **70**, 5810-5817.

Eerlingen, R. C., Cillen, G., & Delcour, J. A. (1994a). Enzyme-resistant starch .4. Effect of endogenous lipids and added sodium dodecyl-sulfate on formation of resistant starch. *Cereal Chemistry*, **71**, 170-177.

Eerlingen, R. C., Crombez, M., & Delcour, J. A. (1993). Enzyme-resistant starch .1. Quantitative and qualitative influence of incubation-time and temperature of autoclaved starch on resistant starch formation. *Cereal Chemistry*, **70**, 339-344.

Eerlingen, R. C., & Delcour, J. A. (1995). Formation, analysis, structure and properties of type-III enzyme resistant starch. *Journal of Cereal Science*, **22**, 129-138.

Eerlingen, R. C., Vandenbroeck, I., Delcour, J. A., Slade, L., & Levine, H. (1994b). Enzyme-resistant starch .6. Influence of sugars on resistant starch formation. *Cereal Chemistry*, **71**, 472-476.

Escarpa, A., Gonzalez, M. C., Morales, M. D., & SauraCalixto, F. (1997). An approach to the influence of nutrients and other food constituents on resistant starch formation. *Food Chemistry*, **60**, 527-532.

Faisant, N., Buleon, A., Colonna, P., Molis, C., Lartigue, S., Galmiche, J. P., & Champ, M. (1995). Digestion of raw banana starch in the small-intestine of healthy humans - structural features of resistant starch. *British Journal of Nutrition*, **73**, 111-123.

Fassler, C., Arrigoni, E., Venema, K., Brouns, F., & Amado, R. (2006a). *In vitro* fermentability of differently digested resistant starch preparations. *Molecular Nutrition & Food Research*, **50**, 1220-1228.

Fassler, C., Arrigoni, E., Venema, K., Hafner, V., Brouns, F., & Amado, R. (2006b). Digestibility of resistant starch containing preparations using two *in vitro* models. *European Journal of Nutrition*, **45**, 445-453.

Ferguson, L. R., Tasman-Jones, C., Englyst, H., & Harris, P. J. (2000). Comparative effects of three resistant starch preparations on transit time and short-chain fatty acid production in rats. *Nutrition and Cancer-an International Journal*, **36**, 230-237.

Flint, H. J., Duncan, S. H., Scott, K. P., & Louis, P. (2007). Interactions and competition within the microbial community of the human colon: links between diet and health. *Environmental Microbiology*, **9**, 1101-1111.

Franco, C. M. L., Ciacco, C. F., & Tavares, D. Q. (1995). Effect of heat-moisture treatment on the enzymatic susceptibility of corn starch granules. *Starch-Starke*, **47**, 223-228.

Gidley, M. J. (1987). Factors affecting the crystalline type (A-C) of native starches and model compounds - a rationalization of observed effects in terms of polymorphic structures. *Carbohydrate Research*, **161**, 301-304.

Gidley, M. J., & Bociek, S. M. (1988). ^{13}C CP/MAS NMR-studies of amylose inclusion complexes, cyclodextrins, and the amorphous phase of starch granules - relationships between glycosidic linkage conformation and solid-state ^{13}C chemical-shifts. *Journal of the American Chemical Society*, **110**, 3820-3829.

Gidley, M. J., & Bulpin, P. V. (1987). Crystallization of malto-oligosaccharides as models of the crystalline forms of starch - minimum chain-length requirement for the formation of double helices. *Carbohydrate Research*, **161**, 291-300.

Gidley, M. J., Cooke, D., Darke, A. H., Hoffmann, R. A., Russell, A. L., & Greenwell, P. (1995). Molecular order and structure in enzyme-resistant retrograded starch. *Carbohydrate Polymers*, **28**, 23-31.

Godet, M. C., Bizot, H., & Buleon, A. (1995). Crystallization of amylose-fatty acid complexes prepared with different amylose chain lengths. *Carbohydrate Polymers*, **27**, 47-52.

Goni, I., Garcia-Alonso, A., Martin-Carron, N., & Saura-Calixto, F. (2000). *In vitro* fermentation of different types of alpha-amylase resistant corn starches. *European Food Research and Technology*, **211**, 316-321.

Goni, I., GarciaAlonso, A., & SauraCalixto, F. (1997). A starch hydrolysis procedure to estimate glycemic index. *Nutrition Research*, **17**, 427-437.

Gunaratne, A., & Hoover, R. (2002). Effect of heat-moisture treatment on the structure and physicochemical properties of tuber and root starches. *Carbohydrate Polymers*, **49**, 425-437.

Hamer, H. M., Jonkers, D., Venema, K., Vanhoutvin, S., Troost, F. J., & Brummer, R. J. (2008). Review article: the role of butyrate on colonic function. *Alimentary Pharmacology & Therapeutics*, **27**, 104-119.

Haralampu, S. G. (2000). Resistant starch - a review of the physical properties and biological impact of RS3. *Carbohydrate Polymers*, **41**, 285-292.

Hizukuri, S. (1986). Polymodal distribution of the chain lengths of amylopectins, and its significance. *Carbohydrate Research*, **147**, 342-347.

Hoover, R. (2001). Composition, molecular structure, and physicochemical properties of tuber and root starches: a review. *Carbohydrate Polymers*, **45**, 253-267.

Htoon, A., Shrestha, A. K., Flanagan, B. M., Lopez-Rubio, A., Bird, A. R., Gilbert, E. P., & Gidley, M. J. (2009). Effects of processing high amylose maize starches under controlled conditions on structural organisation and amylase digestibility. *Carbohydrate Polymers*, **75**, 236-245.

Imberty, A., Buleon, A., Tran, V., & Perez, S. (1991). Recent advances in knowledge of starch structure. *Starch-Starke*, **43**, 375-384.

Jacobasch, G., Dongowski, G., Schmiedl, D., & Muller-Schmehl, K. (2006). Hydrothermal treatment of Novelose 330 results in high yield of resistant starch type 3 with beneficial prebiotic properties and decreased secondary bile acid formation in rats. *British Journal of Nutrition*, **95**, 1063-1074.

Jacobs, H., & Delcour, J. A. (1998). Hydrothermal modifications of granular starch, with retention of the granular structure: a review. *Journal of Agricultural and Food Chemistry*, **46**, 2895-2905.

Jagannath, J. H., Nanjappa, C., Das Gupta, D. K., & Arya, S. S. (2001). Crystallization kinetics of precooked potato starch under different drying conditions (methods). *Food Chemistry*, **75**, 281-286.

Jayakody, L., & Hoover, R. (2008). Effect of annealing on the molecular structure and physicochemical properties of starches from different botanical origins - a review. *Carbohydrate Polymers*, **74**, 691-703.

Jenkins, P. J., & Donald, A. M. (1997). Breakdown of crystal structure in potato starch during gelatinization. *Journal of Applied Polymer Science*, **66**, 225-232.

Kizil, R., Irudayaraj, J., & Seetharaman, K. (2002). Characterization of irradiated starches by using FT-Raman and FT-IR spectroscopy. *Journal of Agricultural and Food Chemistry*, **50**, 3912-3918.

Le Bail, P., Rondeau, C., & Buleon, A. (2005). Structural investigation of amylose complexes with small ligands: helical conformation, crystalline structure and thermostability. *International Journal of Biological Macromolecules*, **35**, 1-7.

Lehmann, U., Jacobasch, G., & Schmiedl, D. (2002). Characterization of resistant starch type III from banana (Musa acuminata). *Journal of Agricultural and Food Chemistry*, **50**, 5236-5240.

Leloup, V. M., Colonna, P., & Ring, S. G. (1991). α-amylase adsorption on starch crystallites. *Biotechnology and Bioengineering*, **38**, 127-134.

Leloup, V. M., Colonna, P., Ring, S. G., Roberts, K., & Wells, B. (1992).

Microstructure of amylose gels. *Carbohydrate Polymers*, **18**, 189-197.

Lesmes, U., Beards, E. J., Gibson, G. R., Tuohy, K. M., & Shimoni, E. (2008). Effects of resistant starch type III polymorphs on human colon microbiota and short chain fatty acids in human gut models. *Journal of Agricultural and Food Chemistry*, **56**, 5415-5421.

Lim, C. C., Ferguson, L. R., & Tannock, G. W. (2005). Dietary fibres as "prebiotics": Implications for colorectal cancer. *Molecular Nutrition & Food Research*, **49**, 609-619.

Long, Y., Shanks, R. A., & Stachurski, Z. H. (1995). Kinetics of polymer crystallization. *Progress in Polymer Science*, **20**, 651-701.

Lopez-Rubio, A., Flanagan, B. M., Gilbert, E. P., & Gidley, M. J. (2008a). A novel approach for calculating starch crystallinity and its correlation with double helix content: a combined XRD and NMR study. *Biopolymers*, **89**, 761-768.

Lopez-Rubio, A., Flanagan, B. M., Shrestha, A. K., Gidley, M. J., & Gilbert, E. P. (2008b). Molecular rearrangement of starch during *in vitro* digestion: Toward a better understanding of enzyme resistant starch formation in processed starches. *Biomacromolecules*, **9**, 1951-1958.

Lopez-Rubio, A., Htoon, A., & Gilbert, E. P. (2007). Influence of extrusion and digestion on the nanostructure of high-amylose maize starch. *Biomacromolecules*, **8**, 1564-1572.

Louis, P., & Flint, H. J. (2009). Diversity, metabolism and microbial ecology of butyrate-producing bacteria from the human large intestine. *FEMS Microbiology Letters*, **294**, 1-8.

Louis, P., Scott, K. P., Duncan, S. H., & Flint, H. J. (2007). Understanding the effects of diet on bacterial metabolism in the large intestine. *Journal of Applied Microbiology*, **102**, 1197-1208.

Macfarlane, G. T., & Gibson, G. R. (1997). Carbohydrate fermentation, energy transduction and gas metabolism in the human large intestine. In: R. I. Mackie, & B. A. White, *Gastrointestinal Microbiology* vol. 1 (pp. 269-318). London: Chapman and Hall.

Macfarlane, G. T., & Macfarlane, S. (1993). Factors Affecting Fermentation Reactions in the Large-Bowel. *Proceedings of the Nutrition Society*, **52**, 367-373.

Macfarlane, G. T., Steed, H., & Macfarlane, S. (2008). Bacterial metabolism and health-related effects of galacto-oligosaccharides and other prebiotics. *Journal of Applied Microbiology*, **104**, 305-344.

Macfarlane, S., & Macfarlane, G. T. (2003). Regulation of short-chain fatty acid production. *Proceedings of the Nutrition Society*, **62**, 67-72.

Martin, L. J. M., Dumon, H. J. W., & Champ, M. M. J. (1998). Production of short-chain fatty acids from resistant starch in a pig model. *Journal of the Science of Food and Agriculture*, **77**, 71-80.

Martin, L. J. M., Dumon, H. J. W., Lecannu, G., & Champ, M. M. J. (2000). Potato and high-amylose maize starches are not equivalent producers of butyrate for the colonic mucosa. *British Journal of Nutrition*, **84**, 689-696.

Miller, T. L., & Wolin, M. J. (1996). Pathways of acetate, propionate, and butyrate formation by the human fecal microbial flora. *Applied Enviromental Microbiology*, **62**, 1589-1592.

Morgan, K. R., Furneaux, R. H., & Larsen, N. G. (1995). Solid-state NMR-studies on the structure of starch granules. *Carbohydrate Research*, **276**, 387-399.

Morrison, D. J., Mackay, W. G., Edwards, C. A., Preston, T., Dodson, B., & Weaver, L. T. (2006). Butyrate production from oligofructose fermentation by the human faecal flora: what is the contribution of extracellular acetate and lactate? *British Journal of Nutrition*, **96**, 570-577.

Mutungi, C., Onyango, C., Doert, T., Paasch, S., Thiele, S., Machill, S., Jaros, D., & Rohm, H. (2011a). Long- and short-range structural changes of recrystallised cassava starch subjected to *in vitro* digestion. *Food Hydrocolloids*, **25**, 477-485.

Mutungi, C., Onyango, C., Jaros, D., Henle, T., & Rohm, H. (2009a). Determination of optimum conditions for enzymatic debranching of cassava starch and synthesis of resistant starch type III using central composite rotatable design. *Starch-Starke*, **61**, 367-376.

Mutungi, C., Onyango, C., Rost, F., Doert, T., Jaros, D., & Rohm, H. (2010). Structural and physicochemical properties and *in vitro* digestibility of recrystallized linear alpha-D-(1→ 4) glucans derived from mild-acid-modified cassava starch. *Food Research International*, **43**, 1144-1154.

Mutungi, C., Passauer, L., Onyango, C., Jaros D., & Rohm, H. (2011b). Debranched cassava starch crystallinity determination by Raman spectroscopy: correlation of features in Raman spectra with X-ray diffraction and ^{13}C CP/MAS NMR spectroscopy. *Carbohydrate Polymers*, doi: 10.1016/j.carbpol.2011.08.032.

Mutungi, C., Rost, F., Onyango, C., Jaros, D., & Rohm, H. (2009b). Crystallinity, thermal and morphological characteristics of resistant starch type III produced by hydrothermal treatment of debranched cassava starch. *Starch-Starke*, **61**, 634-645.

Mutungi, C., Shuldt, S., Onyango, C., Schneider, Y., Jaros, D., & Rohm, H. (2011c). Dynamic moisture sorption characteristics of enzyme-resistant recrystallized cassava starch. *Biomacromolecules*, **12**, 660-671.

Nakamura, Y. (1996). Some properties of starch debranching enzymes and their possible role in amylopectin biosynthesis. *Plant Science*, **121**, 1-18.

Nobakhti, A. R., Ehsani, M. R., Mousavi, S. M., & Mortazavian, A. M. (2009). Influence of lactulose and Hi-maize addition on viability of probiotic microorganisms in freshly made synbiotic fermented milk drink. *Milchwissenschaft-Milk Science International*, **64**, 191-193.

Onyango, C., & Mutungi, C. (2008). Synthesis and *in vitro* digestion of resistant starch type III from enzymatically hydrolysed cassava starch. *International Journal of Food Science and Technology*, **43**, 1860-1865.

Paris, M., Bizot, H., Emery, J., Buzare, J. Y., & Buleon, A. (1999). Crystallinity and structuring role of water in native and recrystallized starches by ^{13}C CP-MAS NMR spectroscopy 1: Spectral decomposition. *Carbohydrate Polymers*, **39**, 327-339.

Paris, M., Bizot, H., Emery, J., Buzare, J. Y., & Buleon, A. (2001). NMR local range investigations in amorphous starchy substrates I. Structural heterogeneity probed by ^{13}C CP-MAS NMR. *International Journal of Biological Macromolecules*, **29**, 127-136.

Perez, S., Imberty, A., & Raymond, P. (1990). Modelling of interactions of polysaccharide chains: Application to crystalline polymorphism of starch granules. In: D. A. French, & J. W. Brandy, *Computer modeling of carbohydrate molecules* (pp. 281-299). Washington, DC: ACS Publications.

Pohu, A., Putaux, J. L., Planchot, V., Colonna, P., & Buleon, A. (2004). Origin of the limited alpha-amylolysis of debranched maltodextrins crystallized in the A form: a TEM study on model substrates. *Biomacromolecules*, **5**, 119-125.

Popov, D., Buléon, A., Burghammer, M., Chanzy, H., Montesanti, N., Putaux, J. L., Potocki-Véronèse, G., & Riekel, C. (2009). Crystal structure of A-amylose: A revisit from synchrotron micro-diffraction analysis of single crystals. *Macromolecules*, **42**, 1167-1174.

Pryde, S. E., Duncan, S. H., Hold, G. L., Stewart, C. S., & Flint, H. J. (2002). The microbiology of butyrate formation in the human colon. *FEMS Microbiology Letters*, **217**, 133-139.

Raben, A., Tagliabue, A., Christensen, N. J., Madsen, J., Holst, J. J., & Astrup, A. (1994). Resistant Starch - the effect on postprandial glycemia, hormonal response, and satiety. *American Journal of Clinical Nutrition*, **60**, 544-551.

Rabiu, B. A., & Gibson, G. R. (2002). Carbohydrates: a limit on bacterial diversity within the colon. *Biological Reviews*, **77**, 443-453.

Ramesh, M., Ali, S. Z., & Bhattacharya, K. R. (1999). Structure of rice starch and its relation to cooked-rice texture. *Carbohydrate Polymers*, **38**, 337-347.

Reader, D., Johnson, M. L., Hollander, P., & Franz, M. (1997). The glycemic and insulinemic response of resistant starch in a food bar vs. two commercially available food bars in persons with type II diabetes mellitus. *Diabetes*, **46**, 975-975.

Reid, C. A., Hillman, K., & Henderson, C. (1998). Effect of retrogradation, pancreatin digestion and amylose/amylopectin ratio on the fermentation of starch by *Clostridium butyricum* (NCIMB 7423). *Journal of the Science of*

Food and Agriculture, **76**, 221-225.

Rideout, T. C., Liu, Q., Wood, P., & Fan, M. Z. (2008). Nutrient utilisation and intestinal fermentation are differentially affected by the consumption of resistant starch varieties and conventional fibres in pigs. *British Journal of Nutrition*, **99**, 984-992.

Robertson, J. A., Ryden, P., Botham, R. L., Reading, S., Gibson, G., & Ring, S. G. (2001). Structural properties of diet-derived polysaccharides and their influence on butyrate production during fermentation. *Food Science and Technology*, **34**, 567-573.

Robyt, J. F. (2009). Enzymes and their action on starch. In: J. BeMiller, & R. Whistler, *Starch: Chemistry and Technology* (pp. 238-292). Amsterdam: Elsevier.

Sajilata, M. G., Singhal, R. S., & Kulkarni, P. R. (2006). Resistant starch - A review. *Comprehensive Reviews in Food Science and Food Safety*, **5**, 1-17.

Schuster, K. C., Ehmoser, H., Gapes, J. R., & Lendl, B. (2000). On-line FT-Raman spectroscopic monitoring of starch gelatinisation and enzyme catalysed starch hydrolysis. *Vibrational Spectroscopy*, **22**, 181-190.

Schwiertz, A., Lehmann, U., Jacobasch, G., & Blaut, M. (2002). Influence of resistant starch on the SCFA production and cell counts of butyrate-producing *Eubacterium* spp. in the human intestine. *Journal of Applied Microbiology*, **93**, 157-162.

Sekirov, I., Russell, S. L., Antunes, L. C. M., & Finlay, B. B. (2010). Gut microbiota in health and disease. *Physiological Reviews*, **90**, 859-904.

Shamai, K., Bianco-Peled, H., & Shimoni, E. (2003). Polymorphism of resistant starch type III. *Carbohydrate Polymers*, **54**, 363-369.

Shibanuma, K., Takeda, Y., Hizukuri, S., & Shibata, S. (1994). Molecular-structures of some wheat starches. *Carbohydrate Polymers*, **25**, 111-116.

Sievert, D., Czuchajowska, Z., & Pomeranz, Y. (1991). Enzyme-resistant starch. III. X-ray-diffraction of autoclaved amylomaize VII starch and enzyme-resistant starch residues. *Cereal Chemistry*, **68**, 86-91.

Silverio, J., Fredriksson, H., Andersson, R., Eliasson, A. C., & Aman, P. (2000). The effect of temperature cycling on the amylopectin retrogradation of starches with different amylopectin unit-chain length distribution. *Carbohydrate Polymers*, **42**, 175-184.

Silvi, S., Rumney, C. J., Cresci, A., & Rowland, I. R. (1999). Resistant starch modifies gut microflora and microbial metabolism in human flora-associated rats inoculated with faeces from Italian and UK donors. *Journal of Applied Microbiology*, **86**, 521-530.

Singh, J., Dartois, A., & Kaur, L. (2010). Starch digestibility in food matrix: a review. *Trends in Food Science & Technology*, **21**, 168-180.

Smits, A. L. M., Kruiskamp, P. H., van Soest, J. J. G., & Vliegenthart, J. F. G. (2003). Interaction between dry starch and plasticisers glycerol or ethylene glycol, measured by differential scanning calorimetry and solid state NMR spectroscopy. *Carbohydrate Polymers*, **53**, 409-416.

Snape, C. E., Morrison, W. R., Maroto-Valer, M. M., Karkalas, J., & Pethrick, R. A. (1998). Solid state ^{13}C NMR investigation of lipid ligands in V-amylose inclusion complexes. *Carbohydrate Polymers*, **36**, 225-237.

Tahir, R., Ellis, P. R., Bogracheva, T. Y., Taylor-Maeres, C., & Butterworth, P. J. (2011). Study of the structure and properties of native and hydrothermally processed wild-type, *lam* and *r* variant pea starches that affect amylolysis of these starches. *Biomacromolecules*, **12**, 123.

Takahashi, Y., Kumano, T., & Nishikawa, S. (2004). Crystal structure of B-amylose. *Macromolecules*, **37**, 6827-6832.

Takata, H., Kuriki, T., Okada, S., Takesada, Y., Iizuka, M., Minamiura, N., & Imanaka, T. (1992). Action of neopullulanase - Neopullulanase catalyzes both hydrolysis and transglycosylation at α-(1 $\rightarrow$ 4)-glucosidic and α-(1 $\rightarrow$ 6)-glucosidic linkages. *Journal of Biological Chemistry*, **267**, 18447-18452.

Tan, I., Flanagan, B. M., Halley, P. J., Whittaker, A. K., & Gidley, M. J. (2007). A method for estimating the nature and relative proportions of amorphous, single, and double-helical components in starch granules by ^{13}C- CP/MAS NMR. *Biomacromolecules*, **8**, 885-891.

Tester, R. F., & Debon, S. J. J. (2000). Annealing of starch - a review. *International Journal of Biological Macromolecules*, **27**, 1-12.

Tester, R. F., Karkalas, J., & Qi, X. (2004). Starch - composition, fine structure and architecture. *Journal of Cereal Science*, **39**, 151-165.

Therien-Aubin, H., & Zhu, X. X. (2009). NMR spectroscopy and imaging studies of pharmaceutical tablets made of starch. *Carbohydrate Polymers*, **75**, 369-379.

Topping, D. L., Bajka, B. H., Bird, A. R., Clarke, J. M., Cobiac, L., Conlon, M. A., Morell, M. K., & Toden, S. (2008). Resistant starches as a vehicle for delivering health benefits to the human large bowel. *Microbial Ecology in Health and Disease*, **20**, 103 -108.

Topping, D. L., & Clifton, P. M. (2001). Short-chain fatty acids and human colonic function: Roles of resistant starch and nonstarch polysaccharides. *Physiological Reviews*, **81**, 1031-1064.

Topping, D. L., Fukushima, M., & Bird, A. R. (2003). Resistant starch as a prebiotic and synbiotic: state of the art. *Proceedings of the Nutrition Society*, **62**, 171-176.

van der Maarel, M. J. E. C., van der Veen, B., Uitdehaag, J. C. M., Leemhuis, H., & Dijkhuizen, L. (2002). Properties and applications of starch-converting enzymes of the α-amylase family. *Journal of Biotechnology*, **94**, 137-155.

van Soest, J. J. G., Tournois, H., deWit, D., & Vliegenthart, J. F. G. (1995). Short-range structure in (partially) crystalline potato starch determined with attenuated total reflectance Fourier-transform IR spectroscopy. *Carbohydrate Research*, **279**, 201-214.

Vermeylen, R., Goderis, B., & Delcour, J. A. (2006). An X-ray study of hydrothermally treated potato starch. *Carbohydrate Polymers*, **64**, 364-375.

Whittam, M. A., Noel, T. R., & Ring, S. G. (1990). Melting behavior of A-type and B-type crystalline starch. *International Journal of Biological Macromolecules*, **12**, 359-362.

Wronkowska, M., Soral-Smietana, M., & Biedrzycka, E. (2008). Utilization of resistant starch of native tapioca, corn and waxy corn starches and their retrograded preparations by *Bifidobacterium*. *International Journal of Food Sciences and Nutrition*, **59**, 80-87.

Wu, H. C. H., & Sarko, A. (1978a). Packing analysis of carbohydrates and polysaccharides .8. Double-helical molecular-structure of crystalline B-amylose. *Carbohydrate Research*, **61**, 7-25.

Wu, H. C. H., & Sarko, A. (1978b). Packing analysis of carbohydrates and polysaccharides .9. Double-helical molecular-structure of crystalline A-amylose. *Carbohydrate Research*, **61**, 27-40.

Zampa, A., Silvi, S., Fabiani, R., Morozzi, G., Orpianesi, C., & Cresci, A. (2004). Effects of different digestible carbohydrates on bile acid metabolism and SCFA production by human gut micro-flora grown in an *in vitro* semi-continuous culture. *Anaerobe*, **10**, 19-26.

List of publications

Publications in peer review journals

Mutungi, C., Passauer, L., Onyango, C., Jaros D., & Rohm, H. (2011). Debranched cassava starch crystallinity determination by Raman spectroscopy: correlation of features in Raman spectra with X-ray diffraction and ^{13}C CP/MAS NMR spectroscopy. *Carbohydrate Polymers*, doi: 10.1016/j.carbpol.2011.08.032.

Mutungi, C., Shuldt, S., Onyango, C., Schneider, Y., Jaros, D., & Rohm, H. (2011). Dynamic moisture sorption characteristics of enzyme-resistant recrystallised cassava starch. *Biomacromolecules*, **12**, 660-671.

Onyango, C., Mutungi, C., Unbehend, G., & Lindhauer, M. G. (2011). Modification of gluten-free sorghum batter and bread using maize, potato, cassava or rice starch. *Lwt-Food Science and Technology*, **44**, 681-686.

Mutungi, C., Onyango, C., Doert, T., Paasch, S., Thiele, S., Machill, S., Jaros, D., & Rohm, H. (2011). Long- and short-range structural changes of recrystallised cassava starch subjected to *in vitro* digestion. *Food Hydrocolloids*, **25**, 477-485.

Onyango, C., Mutungi, C., Unbehend, G., & Lindhauer, M. G. (2011). Rheological and textural properties of sorghum-based formulations modified with variable amounts of native or pregelatinised cassava starch. *Lwt-Food Science and Technology*, **44**, 687-693.

Onyango, C., Mutungi, C., Unbehend, G., & Lindhauer, M. G. (2010). Batter rheology and bread texture of sorghum-based gluten-free formulations modified with native or pregelatinised cassava starch and alpha-amylase. *International Journal of Food Science and Technology*, **45**, 1228-1235.

Mutungi, C., Onyango, C., Rost, F., Doert, T., Jaros, D., & Rohm, H. (2010). Structural and physicochemical properties and *in vitro* digestibility of recrystallised linear α-D-(1 → 4) glucans derived from mild-acid-modified cassava starch. *Food Research International*, **43**, 1144-1154.

Onyango, C., Mutungi, C., Unbehend, G., & Lindhauer, M. G. (2010). Rheological and baking characteristics of batter and bread prepared from pregelatinised cassava starch and sorghum and modified using microbial transglutaminase. *Journal of Food Engineering*, **97**, 465-470.

Mutungi, C., Rost, F., Onyango, C., Jaros, D., & Rohm, H. (2009). Crystallinity, thermal and morphological characteristics of resistant starch type III produced by hydrothermal treatment of debranched cassava starch. *Starch-Starke*, **61**, 634-645.

Onyango, C., Mutungi, C., Unbehend, G., & Lindhauer, M. G. (2009). Creep-recovery parameters of gluten-free batter and crumb properties of bread prepared from pregelatinised cassava starch, sorghum and selected proteins.

International Journal of Food Science and Technology, **44**, 2493-2499.

Mutungi, C., Onyango, C., Jaros, D., Henle, T., & Rohm, H. (2009). Determination of optimum conditions for enzymatic debranching of cassava starch and synthesis of resistant starch type III using central composite rotatable design. *Starch-Starke*, **61**, 367-376.

Onyango, C., & Mutungi, C. (2008). Synthesis and *in vitro* digestion of resistant starch type III from enzymatically hydrolysed cassava starch. *International Journal of Food Science and Technology*, **43**, 1860-1865.

Mutungi, C., Lamuka, P., Arimi, S., Gathumbi, J., & Onyango, C. (2008). The fate of aflatoxins during processing of maize into *muthokoi* - A traditional Kenyan food. *Food Control*, **19**, 714-721.

Oral presentations, posters and articles in conference proceedings

Mutungi, C., Onyango, C., Jaros, D., & Rohm, H. (2011). Structural characteristics of enzyme resistant starch type III: an example of debranched and hydrothermally treated cassava starch. Presentation, 62nd starch convention, 13-14 April, Detmold, Germany.

Mutungi, C., Schuldt, S., Onyango, C., Schneider, Y., Jaros, D., & Rohm, H. (2011). Structure elucidation of polymeric materials by dynamic vapour sorption: example starch. Presentation, TA Instruments user-seminar, 24-25 March, Würzburg, Germany.

Onyango, C., & Mutungi, C. (2010). Rheological and textural properties of gluten-free sorghum bread modified using selected starches. Presentation Humboldt Kolleg Conference, 24-26 November, Nairobi, Kenya.

Mutungi, C., Onyango, C., Jaros, D., & Rohm, H. (2009). Structural characteristics of recrystallised linear α-D-(1→4) glucans derived from mild-acid modified starch. Poster, *European Federation of Food Science and Technology* conference, 11-13 November, Budapest, Hungary.

Mutungi, C., Rost, F., Onyango, C., Jaros, D., & Rohm, H. (2009). Structural properties of resistant starch type III from debranched cassava (*Manihot esculenta* Crantz) starch. Presentation, German Society of Food Technology Conference, 22-24 October, Lemgo, Germany.

Onyango, C., Mutungi, C., Unbehend, G., & Lindhauer, M. G. (2009). Effect of selected proteins on the dough rheology and crumb character of gluten-free bread made from cassava starch and sorghum. Poster, German Society of Food Technology Conference, 22-24 October, Lemgo, Germany.

Mutungi, C., Rost, F., Onyango, C., Jaros, D., & Rohm, H. (2009). Structural properties of resistant starch type III from debranched cassava (*Manihot esculenta* Crantz) starch. Article conference proceedings, German Society of Food Technology Conference, 22-24 October, Lemgo, Germany.

Onyango, C., Mutungi, C., Unbehend, G., & Lindhauer, M. G. (2009). Utilisation of hydrocolloids and emulsifiers to modify the rheological character and crumb quality of gluten-free bread made from pregelatinised cassava starch and sorghum. Poster; German Society of Food Technology Conference, 22-24 October, Lemgo, Germany.

Mutungi, C., Onyango, C., & Rohm, H. (2008). Optimisation of synthesis and functional properties of resistant starch type III from cassava starch. Poster, German Academic Exchange Service conference, 21-22 November, Cologne, Germany.

Onyango, C., & Mutungi, C. (2007). Optimisation of enzyme mediated synthesis of resistant starch type III from cassava starch. Presentation, International Foundation for Science conference, May 28-June 02, Pretoria, South Africa.

Mutungi, C., & Onyango, C. (2007). Integrated post harvest decontamination of aflatoxins – the case of indigenous maize processing methods in Kenya. Article in conference proceedings, Food Science and Nutrition Network of Africa conference on application of chemistry in food science and nutrition for improved food and nutrition security in Africa, 09-11 April, Ouagadougou, Burkina Faso.

Mutungi, C. & Onyango, C. (2007). Integrated post harvest decontamination of aflatoxins – the case of indigenous maize processing methods in Kenya. Poster, Food Science and Nutrition Network of Africa conference on application of chemistry in food science and nutrition for improved food and nutrition security in Africa, 09-11 April, Ouagadougou, Burkina Faso.

Onyango, C. & Mutungi, C. (2006). Strategies for the synthesis of resistant starch type III from cassava starch. Presentation, International Foundation for Science conference on Natural Resources: Chemical, Biological and Environmental Aspects, 10-14 December, Nairobi, Kenya.

Onyango, C. & Mutungi, C. (2006). Breaking out of the maize-induced famine cycles in Kenya: Evaluation of post-harvest processing technologies of alternative carbohydrate foods. Presentation, Resources: Livelihood management reforms and processes of structural change conference, 18-23 September, Gobabeb, Namibia.

Mutungi, C. & Guantai, S. (2006). Mycotoxins in food grains: Effects on human health and practical strategies for control. Presentation, Aflatoxin prevention campaign Ministry of Agriculture, 21 August, Machakos, Kenya.

List of appendices

Appendix 1

Onyango, C., & Mutungi, C. (2008). Synthesis and *in vitro* digestion of resistant starch type III from enzymatically hydrolysed cassava starch. *International Journal of Food Science and Technology*, **43**, 1860-1865.

Appendix 2

Mutungi, C., Onyango, C., Jaros, D., Henle, T., & Rohm, H. (2009). Determination of optimum conditions for enzymatic debranching of cassava starch and synthesis of resistant starch type III using central composite rotatable design. *Starch-Starke*, **61**, 367-376.

Appendix 3

Mutungi, C., Rost, F., Onyango, C., Jaros, D., & Rohm, H. (2009). Crystallinity, thermal and morphological characteristics of resistant starch type III produced by hydrothermal treatment of debranched cassava starch. *Starch-Starke*, **61**, 634-645.

Appendix 4

Mutungi, C., Onyango, C., Rost, F., Doert, T., Jaros, D., & Rohm, H. (2010). Structural and physicochemical properties and *in vitro* digestibility of recrystallised linear α-D-(1→ 4) glucans derived from mild-acid-modified cassava starch. *Food Research International*, **43**, 1144-1154.

Appendix 5

Mutungi, C., Onyango, C., Doert, T., Paasch, S., Thiele, S., Machill, S., Jaros, D., & Rohm, H. (2011). Long- and short-range structural changes of recrystallised cassava starch subjected to *in vitro* digestion. *Food Hydrocolloids*, **25**, 477-485.

Appendix 6

Mutungi, C., Shuldt, S., Onyango, C., Schneider, Y., Jaros, D., & Rohm, H. (2011). Dynamic moisture sorption characteristics of enzyme-resistant recrystallised cassava starch. *Biomacromolecules*, **12**, 660-671.

Appendix 7

Mutungi, C., Passauer, L., Onyango, C., Jaros D., & Rohm, H. (2011). Debranched cassava starch crystallinity determination by Raman spectroscopy: correlation of features in Raman spectra with X-ray diffraction and ^{13}C CP/MAS NMR spectroscopy. *Carbohydrate Polymers*, doi: 10.1016/j.carbpol.2011.08.032.

Appendix 1

Onyango, C. & Mutungi, C. (2008)

Synthesis and *in vitro* digestion of resistant starch type III from enzymatically hydrolysed cassava starch.

International Journal of Food Science and Technology, **43**, 1860-1865.

International Journal of Food Science and Technology 2008, **43**, 1860–1865

Original article

Synthesis and *in vitro* digestion of resistant starch type III from enzymatically hydrolysed cassava starch

Calvin Onyango* & Christopher Mutungi

Kenya Industrial Research and Development Institute, PO Box 30650-00100, Nairobi, Kenya

(Received 9 November 2007; Accepted in revised form 21 April 2008)

Summary Resistant starch type III (RS III) was synthesised from cassava starch by autoclaving followed by debranching with pullulanase, at varied concentrations (0.4–12 U g^{-1}) and times (2–8 h), and recrystallisation (−18 to 90 °C for 1–16 h). The highest RS III yield (22 g/100 g) was obtained at an enzyme concentration of 4 U g^{-1} after 8 h incubation, followed by recrystallisation at 25 °C for 16 h. Varying the recrystallisation conditions indicated that higher RS III yields (30–35 g/100 g) could be obtained at 90 °C within 2 h. Thinning cassava starch using α-amylase prior to debranching using pullulanase did not further increase the RS III content. *In vitro* digestion data showed that whereas 44% RS III was digested after 6 h, the corresponding value for cassava starch was 89%.

Keywords α-Amylase, cassava starch, *in vitro* digestion, pullulanase, resistant starch type III.

Introduction

Resistant starch type III (RS III) is typically quantified as a component of dietary fibre because it resists digestion in the small intestine and instead passes on to the colon where it is fermented by the microflora resulting in the production of short chain fatty acids, gaseous matter and biomass. These compounds confer positive physiological colonic and non-colonic effects when consumed (Xue *et al.*, 1996; Jacobasch *et al.*, 1999; Topping & Clifton, 2001; Topping *et al.*, 2003; Lim *et al.*, 2005; Nugent, 2005). Consequently the importance of RS III as a nutraceutical is rising and it is increasingly being industrially synthesised for use as a food additive (Sajilata *et al.*, 2006). Production of RS III is a two-stage process that begins with hydration and gelatinisation of starch during which amylose is leached from the granules into solution as a random coil polymer. This process is enhanced by acids or enzymes, which partially hydrolyse the starch polymers (Kettlitz *et al.*, 2000; Lehmann *et al.*, 2002, 2003; González-Soto *et al.*, 2004; Onyango *et al.*, 2006). The second stage involves incubation of the material to allow the flexible linear amylose polymers to recrystallise as double helices and form tightly packed helical or spherical structures stabilised by hydrogen bonds. Thus, formation of RS III is a crystallisation process of amylose, into enzyme-resistant double helices stabilised by hydrogen bonds, in a partially crystalline system.

Enzyme-catalysed hydrolysis is preferable to acid-catalysed hydrolysis for synthesis of RS III because the former provide higher yields, significantly improves product quality and reduces energy consumption. Enzymatic methods for the synthesis of RS III involve hydrolysis with isoamylase or pullulanase to eliminate α-D-(1 → 6) glycosidic linkages of amylopectin (Chiu *et al.*, 1994; Kettlitz *et al.*, 2000; Lehmann *et al.*, 2002, 2003; González-Soto *et al.*, 2004). Amylopectin does not favour formation of RS III because its α-D-(1 → 6) glycosidic linkages are hindered in movement and they detract from ordering of amylose (Eerlingen *et al.*, 1994; Thompson, 2000). Contrastingly, amylose favours formation of RS III because it has linear chain polymers whose degree of polymerisation can be optimised. When the degree of polymerisation of amylose is too low (< 10 glucose units) or high (> 35 glucose units) then synthesis of RS III is hindered either because the polymer lacks the minimum chain length required to form crystallites or it cannot reach the required alignment of polymer chains to form resistant crystallites (Schmiedl *et al.*, 2000; Lehmann *et al.*, 2002, 2003). Thus, the aim of this study was to synthesise RS III from cassava starch by eliminating α-D-(1 → 6) glycosidic linkages of amylopectin and partially hydrolysing α-D-(1 → 4) linkages of amylose polymers prior to controlled incubation. The *in vitro* rate of digestion of RS III was determined using pancreatic amylase.

*Correspondent: E-mail: calonyango@yahoo.com

doi:10.1111/j.1365-2621.2008.01764.x

Materials and methods

Cassava starch extraction and composition

Cassava (*Manihot esculenta* Gaerth) tubers were purchased from Kenya Agriculture Research Institute. Starch was extracted from the tubers using pilot scale equipment as described earlier (Onyango *et al.*, 2006). Protein ($n \times 6.25$), moisture, ash, fibre and lipid contents were determined using standard AOAC (2000) methods. Starch and amylose contents were determined as described earlier (Onyango *et al.*, 2006).

Determination of pullulanase activity

Pullulanase activity was determined colorimetrically with pullulan as substrate. The enzyme stock solution (Sigma P-2986; Sigma-Aldrich Chemie GmbH, Steinheim, Germany) was diluted in sodium acetate buffer (20 mM sodium acetate trihydrate in deionised water adjusted to pH 5.0 using 1 M hydrochloric acid) to 10^{-1}, 10^{-2} and 10^{-3} dilutions. A 2% w/v pullulan solution was prepared by adding 2 g pullulan (Sigma P-4516) in 100 mL sodium acetate buffer (pH 5). The solution (50 µL) was diluted further in 350 µL sodium acetate buffer before adding 100 µL pullulanase and the reaction allowed to proceed at 25 °C. The reaction was stopped after exactly 10 min by adding 500 µL copper solution (4 g copper sulphate, 0.185 g sodium sulphate, 23.96 g sodium carbonate, 15.96 g sodium bicarbonate and 12.14 g sodium potassium tartrate dissolved in 1000 mL distilled water), heated in a boiling water bath for 15 min then reacted with 500 µL arsenomolybdate solution (49.43 g ammonium molybdate tetrahydrate, 5.93 g sodium arsenate dibasic heptahydrate and 756 mM sulphuric acid in 1000 mL distilled water). The content of reducing sugars was determined by reading the absorbance using a u.v.-spectrophotometer at 546 nm against glucose standards. Enzyme activity (U mL^{-1}) was expressed as µmoles glucose released per minute of the reaction by 1 mL enzyme solution.

Effect of debranching cassava starch using pullulanase on resistant starch type III content

Cassava starch (5 g) was suspended in 20 mL distilled water and gelatinised by autoclaving at 121 °C for 15 min. The gel was diluted further by adding 20 mL sodium acetate buffer (pH 5) and cooled to 60 °C. Enzyme solutions (2 mL) at different concentrations (1, 2.5, 5, 7.5, 10, 15, 20 and 30 U mL^{-1}) were added to starch suspensions to give treatments equivalent to 0.4, 1, 2, 3, 4, 6, 8 and 12 U g^{-1}. The samples were incubated at 60 °C and 1 mL aliquots taken and analysed for reducing sugars, using 3,5-dinitrosalicyclic acid reagent after 2, 4, 6 and 8 h. The experiment was set as a 8×4 factorial design with two replicates. Analysis of variance was done using GenStat® (VSN International, Hertfordshire, UK) statistical software and the means separated using least significant difference (LSD) test.

Starch hydrolysates treated with different pullulanase concentrations for 8 h and those from the enzyme concentration-treatment time combination giving optimal amount of reducing sugars were used to determine the RS III contents. The samples were heated in a boiling water bath for 10 min to denature the enzyme, cooled and incubated at 25 °C for 16 h then dried in a hot air oven at 60 °C. The samples were ground using a mortar and pestle and the RS III content determined by adding 200 mg sample to 80 mg porcine pancreatic α-amylase (Sigma A-3176) in a 50 mL centrifuge tube and mixing with 10 mL phosphate buffer (pH 6.0). The mixture was incubated at 37 °C for 16 h after which the pH was adjusted to 4.5 using 2% v/v phosphoric acid before adding 200 µL amyloglucosidase (Sigma A-7095). The mixture was incubated at 60 °C for 30 min, centrifuged at 1500 ***g*** for 10 min and the supernatant decanted. The residue was washed with phosphate buffer (pH 7.5) and re-suspended in 10 mL of the same buffer before adding 200 µL protease (Sigma P-2143, 16 mg in 100 mL phosphate buffer at pH 7.5) and incubating at 42 °C for 4 h. Samples were centrifuged at 1500 ***g*** for 10 min, washed twice with 10 mL distilled water and dried in an air oven at 60 °C to constant weight. The content of RS III was determined from the difference in weight of centrifuge tube containing the dry residue and the corresponding tare weight and expressed as a percent of the untreated sample.

Effect of incubation temperature and time on resistant starch type III content

Cassava starch (5 g) was suspended in 20 mL distilled water and gelatinised by autoclaving at 121 °C for 15 min before adding 2 mL pullulanase (10 U mL^{-1}) to give an enzyme concentration of 4 U g^{-1} of starch. The mixture was incubated in a water bath at 60 °C for 8 h after which the samples were heated in a boiling water bath for 10 min to denature the enzyme. The hydrolysates were cooled under running tap water to 25 °C then incubated at −18, 4, 25, 60 and 90 °C for 1, 2, 4, 8 and 16 h. The samples were dried in an air oven at 60 °C to constant weight, ground using a mortar and pestle and the RS III content determined as described earlier. The experiment was set as a 5×5 factorial design with two replicates. Analysis of variance was done using GenStat® statistical software and means separated using LSD test.

Effect of thinning cassava starch prior to debranching

Starch suspensions (10% w/v) were prepared in distilled water and gelatinised by autoclaving at

121 °C for 15 min then cooled under running tap water to 25 °C before mixing with 1 mL enzyme solutions containing 100, 250, 500, 1000 or 2000 U mL^{-1} porcine pancreatic α-amylase to give enzyme concentrations of 20, 50, 100, 200, and 400 U g^{-1} of starch. The starch-enzyme mixture was incubated in a water bath at 37 °C for 15, 30, 60 or 120 min after which it was heated in a boiling water bath to denature the enzyme. A 1 mL aliquot of the mixture was obtained and analysed for reducing sugars using 3,5-dinitrosalicylic acid. The degree of hydrolysis was expressed as mg maltose/100 mg starch. The experiment was set up as a 5 × 4 factorial design with two replicates. Analysis of variance was done using GenStat® statistical software and the means separated using LSD test.

The enzyme concentration-treatment time combinations giving a reducing power of 10 mg maltose/100 mg starch were selected and used to partially hydrolyse cassava starch slurries after which pullulanase (4 U g^{-1}) was added and the mixtures incubated at 60 °C for 1, 2, 4 or 8 h. Recrystallisation was done at 90 °C for 4 h after which the hydrolysates were dried in a hot air oven at 60 °C, ground and the RS III content determined. The experiment was set up as a 3 × 2 × 4 factorial design with two replicates. Analysis of variance was done using GenStat® statistical software and the means separated using LSD test.

In vitro digestibility of cassava starch and resistant starch type III

In vitro digestibility of RS III and cassava starch were determined according to the method of Marlett & Longacre (1996) with modification. The sample (50 mg) was added to 5 mL phosphate buffer (pH 6.9), heated in a boiling water bath for 10 min then cooled to 37 °C. Pancreatin solution was prepared by mixing 1 g porcine pancreatin (Sigma P-1745) with 5 mL phosphate buffer (pH 6.9) containing 0.04% (w/v) sodium chloride for 10 min and the mixture allowed to stand for 10 min before adding 50 μL of the supernatant to the RS III suspension. The mixture was incubated in a shaking water bath at 37 °C for 6 h and samples collected periodically. For each collected sample, enzyme activity was terminated by heating the sample in a boiling water bath for 5 min. The samples were centrifuged at 1500 ***g*** for 10 min, the pellet washed twice with 80% ethanol (5 mL) and dried to constant weight in a hot air oven set at 60 °C. Per cent hydrolysis was calculated by multiplying the difference in weight of undigested sample and dry pellet by 100 and dividing the product with the weight of the undigested sample. The tests were carried out in duplicate.

Results and discussion

Composition of cassava starch

Moisture, protein, fibre, ash and lipid contents of extracted cassava starch were 9.30%, 1.07%, 1.04%, 0.38% and 0.30% respectively. The starch and amylose contents were 92% and 16.57%, respectively, on dry weight basis. These parameters were within values reported earlier for cassava starch extracted by a similar method (Onyango *et al.*, 2006).

Effect of pullulanase concentration on debranching cassava starch and resistant starch type III production

The amount of reducing sugars liberated in 1 min when pullulanase (1 mL) was reacted with pullulan was 0.65 μmol (measured as glucose), which translated to an enzyme activity of 325 U mL^{-1}. This stock solution was appropriately diluted to obtain enzyme concentrations required for our study. The solids content of the starch suspension was adjusted to the highest feasible solids level (25 g/100 g) in order to facilitate uniform blending of enzyme and starch suspension, subsequent to drying and maximise the RS III yields. Other authors reported optimum RS III yields from starch suspensions with at least 15 g/100 g solids (Chiu *et al.*, 1994; Lehmann *et al.*, 2002, 2003; González-Soto *et al.*, 2004).

The interaction effects of enzyme concentration and hydrolysis time were insignificant ($P > 0.05$), whereas the rate of production of reducing sugars increased with time and enzyme concentration (Fig. 1). A similar pattern has been noted during enzymatic synthesis of RS III from banana starch and indicates that complete debranching of amylopectin with pullulanase is time and

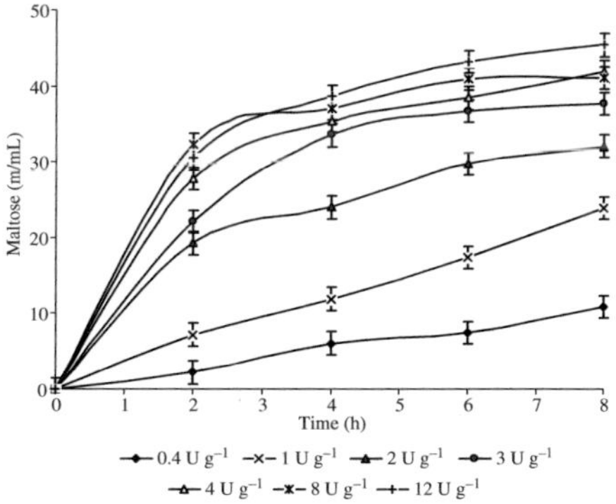

Figure 1 Effect of time and enzyme concentration on production of reducing sugars.

enzyme-concentration dependent (González-Soto *et al.*, 2004). At enzyme concentrations exceeding 3 U g^{-1} the rate of production of reducing sugars tended to level off when the incubation time exceeded 4 h. Debranching starch increases the concentration of linear short chains that contribute to the formation of RS III. Enzyme catalysed hydrolysis is suitable for synthesis of RS III because it is high yielding, significantly improves product quality and reduces energy consumption. Pullulanase is specific to α-D-(1 → 6) glycosidic linkages of amylopectin and partially depolymerises amylopectin chains resulting to increased polymer mobility for molecular rearrangement. The resulting RS III yields ranged from 8.98 to 35.25 g/100 g. Contrastingly, acid hydrolysis of cassava starch gives lower RS III yields (0.59–9.97 g/100 g) because its hydrolytic mode of action is non-specific as it randomly breaks both the α-D-(1 → 4) and α-D-(1 → 6) glycosidic bonds (Onyango *et al.*, 2006). Similarly low RS III yields have been reported from banana starch debranched using acids rather than pullulanase (Lehmann *et al.*, 2003); though these authors finally recommended an integrated acid-enzyme hydrolysis for production of high RS III yields.

The yields of reducing sugars were related to the RS III contents for cassava starch samples hydrolysed for 8 h at different pullulanase concentrations (Fig. 2). Synthesis of RS III correlated positively with increasing yields of reducing sugars up to an enzyme concentration of 4 U g^{-1} after which the RS III content declined while the amount of reducing sugars increased as enzyme concentration increased from 8 to 12 U g^{-1}. Debranching of starch with pullulanase gives rise to linear low molecular weight polymers which promote retrogradation. However, there is also production of oligosaccharides whose chain lengths are not long enough to recrystallise into RS III, and which possibly sterically impeded retrogradation at the high enzyme concentrations. In order to increase the RS III yield, the hydrolysate should be washed with water to remove these interfering substances (Lehmann *et al.*, 2002, 2003).

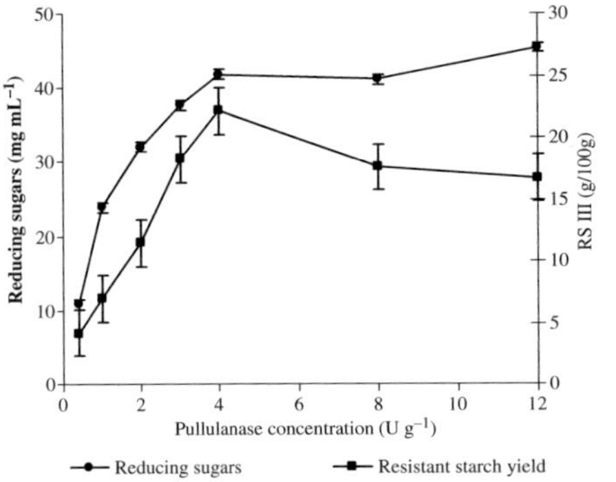

Figure 2 Effect of pullulanase concentration on reducing sugars and resistant starch type III content.

Effect of incubation temperature and time on resistant starch type III production

Synthesis of RS III is influenced by several factors such as nature of starch, amylose content, starch concentration in the gel, hydrolysis and incubation conditions and presence of lipids or low molecular weight substances such as sugars (Chiu *et al.*, 1994; Skrabanja *et al.*, 1999; Schmiedl *et al.*, 2000; Lehmann *et al.*, 2002, 2003; González-Soto *et al.*, 2004). The most important variables during recrystallisation of amylose in a partially crystalline system are incubation temperature and time. We studied the influence of incubation, at various time-temperature combinations, on RS III production from cassava starch hydrolysed for 8 h at a pullulanase concentration of 4 U g^{-1} (Table 1). The interaction effect of incubation time and temperature significantly influenced the RS III yields ($P < 0.05$) with the highest yields being noted in samples incubated at 90 °C for 2–8 h. Generally, higher RS III contents were obtained in shorter time periods when enzyme-mediated hydrolysis was compared to acid-mediated hydrolysis (Onyango *et al.*, 2006), though the yields from enzyme hydrolysed cassava starch could not be easily interpreted on the basis of nucleation, propagation and maturation. However, it appears that nucleation, propagation and maturation all occurred within the first 2 h for hydrolysates incubated at 90 °C.

Resistant starch type III yields from cassava starch incubated at 4 °C did not differ ($P > 0.05$) from those at −18 °C in the first 2 h but thereafter significantly increased ($P < 0.05$) with time whereas samples incubated at −18 °C did not change with time ($P > 0.05$). The RS III contents from cassava starch incubated at intermediate temperature ranges (25 or 60 °C) did not differ

Table 1 Effect of incubation temperature and time on resistant starch type III content*

Incubation temperature (°C)	Incubation time (h) 1	2	4	8	16
−18	14.39^{b}	12.17^{ab}	10.87^{a}	11.52^{a}	8.98^{a}
4	14.52^{bc}	15.89^{cb}	16.04^{c}	19.59^{d}	21.45^{de}
25	22.83^{e}	24.27^{e}	22.00^{e}	23.66^{ef}	25.36^{f}
60	23.35^{e}	23.48^{e}	24.27^{e}	27.78^{fg}	27.14^{f}
90	30.90^{g}	31.44^{gh}	34.06^{h}	35.25^{h}	23.28^{e}

Means sharing similar superscripts along the same row or column are not significantly different at $P = 0.05$.
*g/100 g starch.

significantly ($P > 0.05$) irrespective of the incubation time. Synthesis of RS III is inhibited at low temperatures by the formation of ice crystals within the recrystallising linear chains that limit the mobility and realignment of the chains and result in a pulverous product that is more susceptible to amylolytic attack. Similarly to acid hydrolysed cassava starch incubated at freezing temperatures (Onyango *et al.*, 2006), a large proportion of the RS III was probably synthesised during nucleation with no further increase in yields during maturation and propagation. Starch retrogradation temperature is important as it influences starch crystalline structure and subsequent nutritional and functional properties of RS III. X-ray diffraction analysis of RS III shows that retrogradation at low temperatures favours formation of B-type polymorph whereas mixtures of A-, with V-, or B-type polymorphs are formed at high temperatures (Shamai *et al.*, 2003). These polymorph structures have implications on *in vivo* digestion since RS III with A-type patterns are attacked more rapidly than those with B-type patterns (Alonso *et al.*, 1998).

Effect of thinning cassava starch prior to debranching

The reducing sugars content of cassava starch, hydrolysed using α-amylase, was used to determine the optimum enzyme-time treatment combination required for thinning prior to debranching (Table 2). The interaction effect of α-amylase concentration and incubation time was significant ($P < 0.05$) and the reducing sugars contents increased with increasing levels of these factors. Hydrolysates with reducing sugars contents less than ten facilitate synthesis of RS III (Kettlitz *et al.*, 2000). Consequently gelatinised cassava starch was thinned using α-amylase concentrations of 20, 50 and 100 U g^{-1} and incubated for 15 or 30 min prior to debranching with pullulanase and recrystallisation at 90 °C for 4 h (Table 3). The interaction effect of thinning conditions (α-amylase concentration and treatment time) and debranching time was significant ($P < 0.05$) and RS III content was highest (33.13 g/100 g) when gelatinised cassava starch was thinned using an α-amylase concentration of 20 U g^{-1} for 15 min (Table 3). Generally, RS III yields from cassava starch thinned prior to debranching were not appreciably higher than those from cassava starch treated with pullulanase (Table 1). Thinning enhances the activity of pullulanase on amylopectin since unlike isoamylase it exhibits limited activity towards branched longer chain dextrins (Henrik & Allan, 2001). We had, therefore, anticipated that increased yields of low molecular weight fragments after amylolytic and pullulanase hydrolysis of cassava starch would facilitate recrystallisation and thus increase RS III yields. The degree of polymerisation of the glucose units determines the ease with which RS III can be synthesised. Excessive hydrolysis of amylose gives low RS III contents because the resultant polymers are not long enough to form enzyme-resistant crystallites, whereas insufficient hydrolysis is undesirable since the long chain polymers cannot effectively realign to form enzyme-resistant crystallites (Thompson, 2000).

Table 2 Reducing sugars* contents of cassava starch thinned using α-mylase

Concentration of α-amylase (U g^{-1})	Treatment time (min) 15	30	60	120
20	1.49^a	1.79^a	2.55^b	2.71^b
50	2.17^b	3.21^b	3.33^b	4.56^c
100	7.79^d	13.33^e	12.69^e	16.96^f
200	9.33^d	12.21^e	12.75^e	14.77^f
400	30.44^g	30.61^g	31.64^g	38.71^h

Means sharing similar superscripts along the same row or column are not significantly different at $P = 0.05$.
*mg maltose/100 mg starch.

Table 3 Effect of thinning and debranching cassava starch on resistant starch type III contents*

Concentration of α-amylase (U g^{-1})	Treatment time (min)	Debranching time (h) 1	2	4	8
20	15	14.18^a	24.17^{cd}	33.13^f	29.24^e
	30	19.31^b	25.60^d	23.20^c	26.32^d
50	15	15.76^a	23.89^c	27.64^d	28.21^e
	30	19.47^b	21.89^c	17.84^b	18.90^b
100	15	19.66^b	23.98^c	18.58^b	17.43^b
	30	18.63^b	19.31^b	19.13^b	17.61^b

Means sharing similar superscripts along the same row or column are not significantly different at $P = 0.05$.
*g/100 g starch.

In vitro digestion of cassava starch and resistant starch type III

The course of *in vitro* digestion of cassava starch and its RS III derivative is given in Fig. 3. Resistant starch type III had a low digestion rate as only 31% was digested within 1 h and the amount digested increased to 44% after 6 h ($P > 0.05$). Contrastingly, 74% cassava starch was digested within 1 h and the amount increased to 89% after 6 h. The high digestion rate of cassava starch indicates easy accessibility of amylolytic enzymes to the starch polymers. Contrastingly, RS III is largely inaccessible to enzymatic hydrolysis because the linear α-D-(1 → 4) linkages of amylose polymers have recrystallised as double helices and formed tightly packed helical or spherical structures stabilised by hydrogen bonds. The rate and extent of starch and RS III digestion affects several human physiological functions and consequently health and are therefore regarded as

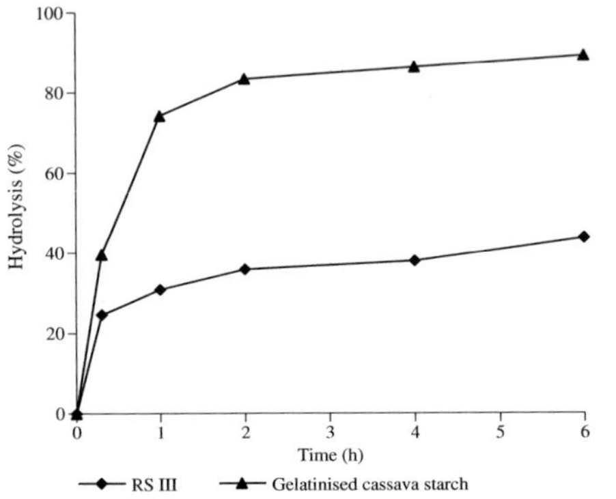

Figure 3 *In vitro* digestion of cassava starch and resistant starch type III.

predictors of metabolic responses to complex carbohydrates *in vitro*. A slow rate of release of glucose in the blood stream results in low glycaemic and insulinaemic responses (Lund & Johnson, 1991; Xue *et al.*, 1996) with associated benefits for weight management.

Conclusion

Resistant starch type III was synthesised from cassava starch after hydrolysis using pullulanase and incubation at controlled conditions. Maximum RS III yields were obtained when the starch was hydrolysed with pullulanase (4 U g^{-1}) for 8 h prior to incubation at 90 °C for 2 h. Thinning of the starch hydrolysate with α-amylase prior to debranching with pullulanase did not further increase the RS III content. *In vitro* digestion of RS III revealed a slow rate of digestion and indicates that the product could be suitable for formulation of foods for diabetics. The economic viability of enzyme-mediated synthesis of RS III from cassava starch is highly promising because the specific nature of enzyme action facilitates synthesis of large quantities of RS III, and quality products that requires minimal downstream processing with minimal energy requirements.

Acknowledgment

Calvin Onyango is grateful to International Foundation for Science (IFS) for the research grant.

References

Alonso, A.G., Calixto, F.S. & Delcour, J.A. (1998). Influence of botanical source and processing on formation of resistant starch type III. *Cereal Chemistry*, **75**, 802–804.

AOAC (2000). In: *Official Methods of Analysis of the Association of Official Analytical Chemists*, 17th edn (edited by W. Horwitz). Gaithersburg, MD: AOAC International.

Chiu, C-W., Henley, M. & Altieri, P. (1994). Process for making amylase resistant starch from high amylose starch. *U.S. Patent*, 5,281,276.

Eerlingen, R.C., Jacobs, H. & Delcour, J.A. (1994). Enzyme-resistant starch. V. Effect of retrogradation of waxy maize starch on enzyme susceptibility. *Cereal Chemistry*, **71**, 351–355.

González-Soto, R.A., Agama-Acevedo, E., Solorza-Feria, J., Rendón-Villalobos, R. & Bello-Pérez, L.A. (2004). Resistant starch made from banana starch by autoclaving and debranching. *Starch/Stärke*, **56**, 495–499.

Henrik, B. & Allan, S. (2001). Starch debranching enzymes. *U.S. Patent*, 6,265,197.

Jacobasch, G., Schmiedl, D., Kruschewski, M. & Schmehl, K. (1999). Dietary resistance starch and inflammatory bowel diseases. *International Journal of Colorectal Diseases*, **14**, 201–211.

Kettlitz, B.W., Coppin, J.V.J.-M., Roper, H.W.W. & Bornet, F. (2000). Highly fermentable resistant starch. *U.S. Patent*, 6,043,229.

Lehmann, U., Jacobasch, G. & Schmiedl, D. (2002). Characterisation of resistant starch type III from banana (*Musa acuminata*). *Journal of Agricultural and Food Chemistry*, **50**, 5236–5240.

Lehmann, U., Rössler, C., Schmiedl, D. & Jacobasch, G. (2003). Production and physicochemical characterisation of resistant starch type III derived from pea starch. *Nahrung/Food*, **47**, 60–63.

Lim, C.C., Ferguson, L.R. & Tannock, G.W. (2005). Dietary fibres as "prebiotics": implications for colorectal cancer. *Molecular Nutrition and Food Research*, **49**, 609–619.

Lund, E.K. & Johnson, I.J. (1991). Fermentable carbohydrate reaching the colon after ingestion of oats in humans. *Journal of Nutrition*, **121**, 311–317.

Marlett, J.A. & Longacre, M.J. (1996). Comparison of *in vitro* and *in vivo* measures of resistant starch in selected grain products. *Cereal Chemistry*, **73**, 63–68.

Nugent, A.P. (2005). Health properties of resistant starch. *Nutrition Bulletin*, **30**, 27–54.

Onyango, C., Bley, T., Jacob, A., Henle, T. & Rohm, H. (2006). Influence of incubation temperature and time on resistant starch type III formation from autoclaved and acid hydrolysed cassava starch. *Carbohydrate Polymers*, **66**, 494–499.

Sajilata, M.G., Singhal, R.S. & Kulkarni, P.R. (2006). Resistant starch: a review. *Comprehensive Reviews in Food Science and Food Safety*, **5**, 1–17.

Schmiedl, D., Bäuerlein, M., Bengs, H. & Jacobash, G. (2000). Production of heat-stable, butyrogenic resistant starch. *Carbohydrate Polymers*, **43**, 183–193.

Shamai, K., Bianco-Peled, H. & Shimoni, E. (2003). Polymorphism of resistant starch type III. *Carbohydrate Polymers*, **54**, 363–369.

Skrabanja, V., Liljeberg, H.G.M., Hedley, C.L., Kreft, I. & Bjork, I.M.E. (1999). Influence of genotype and processing on the *in vitro* rate of starch hydrolysis and resistant starch formation in peas (*Pisum sativum* L). *Journal of Agricultural and Food Chemistry*, **47**, 2033–2039.

Thompson, D.B. (2000). Strategies for the manufacture of resistant starch. *Trends in Food Science and Technology*, **11**, 245–253.

Topping, D.L. & Clifton, P. (2001). Short-chain fatty acids and human colonic function: roles of resistant starch and non-starch polysaccharides. *Physiological Reviews*, **81**, 1031–1064.

Topping, D.L., Fukushima, M. & Bird, A.R. (2003). Resistant starch as a prebiotic and symbiotic: state of the art. *Proceedings of the Nutrition Society*, **62**, 171–176.

Xue, Q., Newman, R.K. & Newman, C.W. (1996). Effects of heat treatment of barley starches on *in vitro* digestibility and glucose responses in rats. *Cereal Chemistry*, **73**, 588–592.

Appendix 2

Mutungi, C., Onyango, C., Jaros, D., Henle, T., & Rohm, H. (2009a)

Determination of optimum conditions for enzymatic debranching of cassava starch and synthesis of resistant starch type III using central composite rotatable design.

Starch-Starke, **61**, 367-376.

Christopher Mutungi[a]
Calvin Onyango[b]
Doris Jaros[a]
Thomas Henle[c]
Harald Rohm[a]

[a] Institute of Food Technology and Bioprocess Engineering, Technische Universität Dresden, Dresden, Germany
[b] Food Technology Division, Kenya Industrial Research and Development Institute, Nairobi, Kenya
[c] Institute of Food Chemistry, Technische Universität Dresden, Dresden, Germany.

Determination of Optimum Conditions for Enzymatic Debranching of Cassava Starch and Synthesis of Resistant Starch Type III using Central Composite Rotatable Design

Cassava starch was debranched by treatment with isoamylase and pullulanase and the yield of resistant starch type III (RS III) optimized with respect to starch solids concentration (7.5-15%, w/v), incubation time (8-24 h) and enzyme concentration using central composite rotatable design. Higher concentrations of pullulanase (10-35 U/g starch) compared to isoamylase (30–90 mU/g starch) were required to give a similar degree of starch hydrolysis within the experimental domain. A clear debranching endpoint was identifiable by following the reducing value, blue value and β-hydrolysis limit of cassava starches debranched using isoamylase. It was difficult to define a debranching endpoint of pullulanase treatment by these parameters due to contaminating α-D-(1→4) activity. The yield of RS III was significantly higher in isoamylolysates and increased steadily with increasing degree of hydrolysis to peak at 57.3%. Purification of the debranched material further increased the RS III yield to 64.1%. Prolonged (24 h) hydrolysis of cassava starch with high concentration of pullulanase (35 U/g) gave lower RS III contents in the purified (34.2%) and unpurified (36.2%) hydrolysates compared to 49.5 and 62.4%, respectively, at moderate pullulanase concentration (22.5 U/g) and incubation time (16 h).

Keywords: Cassava starch; Isoamylase; Pullulanase; Resistant starch type III

1 Introduction

Industrial production of functional retrograded starch, RS III, for use as food ingredient is on the rise due to the physiological and health promoting benefits associated with it. Owing to its ability to withstand food processing conditions and digestion in the upper gut, a large proportion of RS III is fermented in the lower gut into short-chain fatty acids and other health promoting metabolites, besides supporting growth of beneficial gut microflora, and maintaining a lowered colonic pH [1–3]. Optimization of yield and functional properties of RS III are a function of the physicochemical properties of the starting material, its pre- and post-gelatinization modifications, retrogradation conditions and post-retrogradation treatments [4]. Resistant starch type III is formed when linear α-D-(1→4) polyglucans of starch realign themselves and intertwine to form double helices stabilized by hydrogen bonds. The realignment process is enhanced when a large proportion of the starch is present as straight-chain polymers and is impaired by the presence of side branches [4, 5]. Consequently, high-amylose starches are preferred for RS III production whereas synthesis of RS III from low amylose starches is enhanced if preceded by physical, chemical or enzymatic treatments to linearize the amylopectin polymers [6-10].

Cassava starch comprises 76-83% amylopectin and 17–24% amylose with a considerably high degree of branching [11–13]. An average polymer chain length of 20-65 glucose units has been proposed for the production of functional and stable RS III crystallites [9, 14–16]. Whereas the polymer chain-length of cassava amylose and its side branches is considerably large [13, 29], cassava amylopectin comprises short-chain glucans whose ability to form functional RS III crystallites can be increased through depolymerization of the starch with enzymes that selectively hydrolyze the α-D-(1→6) glucosidic linkages [6, 7, 10]. Starch debranching enzymes belong to the α-amylase group of enzymes that hydrolyze or transfer bonds with the retention of the anomeric configuration of the processed bond [17-19]. These are categorized into direct debranching enzymes that directly hydrolyze α-D-(1→6) linkages of α-polyglucans and the indirect debranching enzymes that cleave α-D-

Correspondence: *Christopher Mutungi*, Institute of Food Technology and Bioprocess Engineering, Technische Universität Dresden, 01062 Dresden, Germany. Phone: +49-351-463-36943, e-mail: chrismutungi@yahoo.co.uk.

(1→6) branches by the combined actions of 4-α-glucanotransferase and amylo-1,6-glucosidase. Direct debranching enzymes, pullulanase and isoamylase, are mainly distinguished by the ability of the former to hydrolyze both pullulan and amylopectin and not glycogen which, in addition to amylopectin, is hydrolyzed by the latter [18]. Furthermore, pullulanases show preferential activity against α-limit dextrin, and a retarded activity against large molecular weight glucans, whereas isoamylases attack more readily the large molecular size polymers [18, 19]. Environmental factors that influence the activity of debranching enzymes include temperature, pH and ionic strength whereas substrate specific factors include solids concentration, enzyme concentration and incubation time. Moreover, catalytic productivity of an enzyme preparation is influenced by its purity and the proportion of the enzyme that may be in an inactive or partially active state. The extent of debranching can be monitored by measuring the reducing value of the starch, viscosity, iodine affinity, change in molecular weight or β-hydrolysis limit [7]. In addition, the degree of debranching influences aggregation and recrystallization of linear polymer into microfilaments that constitute RS III. The aim of this study was to investigate α-D-(1→6) glucanohydrolytic effect of pullulanase and isoamylase on cassava starch and the subsequent recrystallization of the debranched starch into RS III.

2 Materials and Methods

2.1 Materials

Cassava starch was obtained from Kenya Industrial Research and Development Institute, Nairobi, Kenya. The extraction method and chemical composition of the starch are described in an earlier publication [6]. Pullulanase from *Bacillus acidopullulyticus* and isoamylase from *Pseudomonas sp.* and were purchased from Sigma-Aldrich, Steinheim, Germany. To determine enzyme activity, a pullulanase stock solution (Sigma P2986) was diluted hundred-fold in 20 mmol/L sodium acetate buffer (pH 5), and 100 μL of the diluted enzyme solution mixed with 50 μL 2% (w/v) pullulan (Sigma P4516) solution [6]. The mixture was incubated at 25°C for 10 min after which reducing sugars were determined colorimetrically using the Nelson-Somogyi alkaline copper reduction procedure [20, 21]. Enzyme activity (U/mL) was expressed as micromoles of glucose released per millilitre enzyme solution within 1 min. Isoamylase (Sigma I5284) was similarly assayed using 2% (w/v) glycogen (Sigma G8751) as substrate in 50 mmol/L sodium acetate buffer, pH 3.5.

2.2 Debranching of cassava starch and determination of degree of hydrolysis

Cassava starch (200 mg) was weighed into 15 mL polypropylene centrifuge tubes and suspended in 3840, 2470, 1580, 1140, or 940 μL of either 20 mmol/L sodium acetate buffer (pH 5.0) or 50 mmol/L sodium acetate buffer (pH 3.5) containing 0.02% (w/v) sodium azide. The starch suspensions were heated in a boiling water bath while mixing to form a homogeneous gel and gelatinized by autoclaving at 121°C for 15 min. The starch gels were cooled to 50°C and mixed with 200 μL of isoamylase or pullulanase diluted to various concentrations (Tab. 1). The starch-enzyme mixtures were carefully mixed using a glass rod and incubated horizontally in a shaking water bath (GGL 1083, Burgwedel, Germany) maintained at 50°C and 170 strokes per minute for 2.5-29.5 h (Tab. 1). The samples were mixed manually every 30 min during the first 2 h of incubation. After incubation, deionized water was added to bring the solids concentrations to 4% (w/v). The mixtures were then heated at 98°C for 20 min to denature the enzyme and stored at 25°C for 16 h.

Subsequent to homogenization, the hydrolyzed starches were analyzed for total soluble carbohydrate content and reducing value. Total soluble carbohydrate content was determined using the phenol-sulphuric acid procedure [22]. A 20 μL aliquot was diluted in 10 mL deionized water, vigorously homogenized and a 100 μL aliquot drawn and diluted in 400 μL deionized water in a separate test-tube. The diluted sample was then mixed with 500 μL of 5% (w/v) phenol prepared in 0.1 mol/L hydrochloric acid after which 2.5 mL 97% (v/v) sulphuric acid was added, stirred on a vortex mixer and allowed to cool to 25 °C before reading absorbance at 490 nm against similarly treated glucose standards containing 0, 10, 20, 30, 40 and 50 μg glucose monohydrate in 500 μL deionized water. The reducing value was determined using the Nelson-Somogyi alkaline copper reduction method [20, 21]. Aliquots (50 μL) of the homogenized sample were added to 450 μL of deionized water, mixed with 500 μL copper solution (4 g copper sulphate, 0.185 g sodium sulphate, 23.96 g sodium carbonate, 15.96 g sodium bicarbonate and 12.14 g sodium potassium tartrate dissolved in 1000 mL distilled water) and heated in a boiling water bath for 60 min. The mixture was cooled to 25°C and reacted with 500 μL arsenomolybdate solution (49.43 g ammonium molybdate tetrahydrate, 5.93 g sodium arsenate dibasic heptahydrate and 756 mmol/L sulphuric acid in 1000 mL distilled water). The content of reducing sugars was determined by reading the absorbance at 546 nm against standards containing 0, 40, 80, 120, 160 and 200 μg glucose monohydrate in 500 μL deionized water. The degree of hydrolysis in percent (DH) was calculated by dividing

Tab. 1. Central composite experimental arrangement for debranching of cassava starch.

Trial	Starch concentration [%]	Enzyme concentration		Incubation time [h]
		Isoamylase [mU/g]	Pullulanase [U/g]	
1	7.5	90	35	8
2	7.5	90	35	24
3	15	90	35	8
4	15	90	35	24
5	7.5	30	10	8
6	7.5	30	10	24
7	15	30	10	8
8	15	30	10	24
9	4.9	60	22.5	16
10	17.6	60	22.5	16
11	11.25	9.5	1.5	16
12	11.25	110	43.5	16
13	11.25	60	22.5	2.5
14	11.25	60	22.5	29.5
15–23	11.25	60	22.5	16

the difference between the reducing value of debranched starch and the reducing value of the enzyme blank by the difference between total carbohydrate content of an equivalent sample and total carbohydrate content of the enzyme blank multiplied by 100.

A rotatable orthogonal central composite design with three factors: starch concentration (7.5-15%, w/v), enzyme concentration (10-35 U/g and 30-90 mU/g for pullulanase and isoamylase, respectively) and incubation time (8-24 h) was used to determine the optimum conditions for the debranching of cassava starch (Tab. 1). The low, middle and high factor levels were coded as −1, 0 and +1, respectively. In order to obtain rotatability, the axial distance for star points (α) was calculated as fourth root of the product of factor levels arranged in a complete factorial design, and consequently the lower and upper star point factor levels coded −1.68 and +1.68, respectively. A total of 23 randomized debranching trials were carried out with each enzyme. The first 8 trials conform to a complete factorial design ($2 \times 2 \times 2$), whereas trials 9–14 conform to the star design. Trials 15–23 represent nine replications of the central points so as to obtain orthogonality and isovariance by rotation properties as well as permit calculation of experimental error variance. For each independent variable, the real experimental values were calculated as total of the product of code value multiplied by half the difference between the high and low level values selected for the variable summed up with the average of the high and low level values of the particular variable. Degree of hydrolysis was the response variable and second-order quadratic models fitted using Statgraphics Centurion XV® professional edition (StatPoint, Inc. Northern Virginia, Washington, D.C., USA). Model coefficients were tested for statistical significance using multiple regression on Instat+v3.36® (Statistical Services Centre, University of Reading, UK) and further random trials at selected variable levels conducted to validate the fitted models.

2.3 Blue value, total soluble carbohydrates and β-hydrolysis limit

The blue value (BV) of starch hydrolysates was determined as described by *Thitipraphunkul* et al. [23] with modification. A suspension of the debranched starch (0.5 mL) was diluted in 9.5 mL deionized water, homogenized and 0.4 mL aliquots were mixed with 3.6 mL dimethyl sulphoxide (DMSO). The starch-DMSO mixture was heated at 90°C for 20 min with occasional stirring to solubilize the starch and diluted to 10 mL with deionized water. The starch content of the solution was approximately 0.8 mg. Iodine solution (160 μL) prepared by dissolving 1 g iodine in 100 mL of 10% (w/v) potassium iodide was added to obtain a starch: iodine: potassium iodide ratio of 1:2:20. The starch-iodine solution was allowed to stand at 25°C for 16 h before reading the absorbance at 680 nm and scanning the wavelength of maximum absorbance (λ_{max}) between 400 and 700 nm. The procedure was repeated after washing the debranched starch twice by suspending 0.5 mL of the homogenate in 9.5 mL deionized water, centrifuging at $1500 \times g$ for 10 min and decanting the supernatant. The pellet was re-suspended in 9.5 mL deionized water after the second washing for the repeat test.

Total soluble carbohydrate content was determined by sampling from the supernatant of centrifuged (1500 × *g* for 10 min) debranched starches as described earlier, whereas β-amylolysis limit was determined using sweet potato β-amylase according to *Tattiyakul* et al. [24] with modification. Aliquots (20 μL) of homogenized debranched starch suspension were mixed with 180 μL DMSO and heated at 90°C for 20 min. The starch-DMSO solution was mixed with 1800 μL β-amylase (Sigma A7005) solution (50 U/mL) prepared in 40 mmol/L sodium acetate buffer (pH 4.8) containing 0.02% (w/v) sodium azide, and incubated in a shaking water bath (37°C, 170 strokes per minute) for 48 h. Aliquots (250 μL) were drawn and analyzed for reducing sugar as glucose following the Nelson-Somogyi alkaline copper reduction procedure [20, 21] and the total carbohydrate content (as glucose) of an equivalent sample was determined following the phenol-sulphuric acid method [22]. Percent β-hydrolysis limit was calculated by dividing the difference between the reducing values of sample and enzyme blank by the difference between total carbohydrate contents of sample and enzyme blank multiplied by 190. Correlations between β-hydrolysis limit, BV, and total soluble carbohydrates with DH were tested using multiple variable analysis on Statgraphics®.

2.4 Resistant starch type III content

Starch hydrolysates were homogenized and partitioned into two equal volumes. In one portion the suspended starch was precipitated by adding an equal volume of 96% (v/v) ethanol, homogenizing and decanting the supernatant after centrifugation at 1500 × *g* for 10 min. The other portion was purified by washing twice with an equal volume of deionized water followed by centrifugation. The pellets were freeze-dried, and RS III contents determined gravimetrically using porcine pancreatic α-amylase and amyloglucosidase. The freeze-dried material (50 mg) was weighed into pre-weighed 15 mL centrifuge tubes, suspended in 5 mL 0.1 mol/L sodium acetate buffer, pH 5.2, containing 0.02% (w/v) sodium azide, vigorously homogenized and mixed with 200 μL enzyme solution containing 2400 U/mL of porcine pancreatic α-amylase and 60 U/mL amyloglucosidase. The enzyme solution was prepared by suspending 1 g porcine pancreatic α-amylase (Sigma A3176) in 0.1 mol/L sodium acetate buffer (pH 5.2) containing 0.02% (w/v) sodium azide and incubated at 37°C in a shaking water bath for 10 min. The suspension was centrifuged at 1500 × *g* for 10 min, the supernatant recovered and mixed with 2.5 mL amyloglucosidase (Sigma A7095) just before use. The starch-enzyme mixtures were incubated horizontally in a shaking water bath (37°C, 170 strokes per minute) for 16 h. To denature the enzymes and precipitate the suspended starch, an equal amount of absolute ethanol (5 mL) was added, the mixture homogenized and centrifuged at 1500 × *g* for 10 min. The pellet was washed with 80% (v/v) ethanol, freeze-dried and weighed after equilibrating to 25°C. The RS III content was determined as a percent ratio of the weight of the residue to the initial weight of sample. Multiple response optimization with DH and RS III content as response variables was performed on Statgraphics® to identify the optimum debranching level for RS III formation.

3 Results and Discussion

3.1 Debranching of cassava starch and determination of degree of hydrolysis

The total carbohydrate content of gelatinized native cassava starch was 967 mg glucose/g with a reducing value of 0.74 mg glucose/g starch which corresponds to a DH of 0.07%. The low reducing value of native cassava starch implies an abundance of undegraded amylopectin and amylose polymers in the starch. The reducing values of debranched cassava starches ranged between 26.3-54.9 mg glucose/g starch, which corresponds to an average degree of polymerization (DP_n) of 17.6-36.7 glucose units and a DH of 2.7–5.3% and 3.4-5.6% for isoamylase and pullulanase, respectively. A paired *t*-test revealed no significant differences for the two sets of DH ($t = 1.21$, $p > 0.05$), implying that the enzymes elicited catalytic effects of similar magnitude on cassava starch at the experimental region within which the debranching was conducted. Cleavage of α-D-(1→6) or α-D-(1→4) glucosidic linkages results in non-combined hemiacetals on the affected bonds which readily convert into reducing aldehydic end groups in alkaline solution. The highest recorded DH of debranched starches was 5.3 and 5.6% for isoamylase and pullulanase, respectively. These values were slightly higher than the number of branched linkages in amylopectin molecules, estimated to be 4-5% of total glucosidic linkages [25] This implies the presence of either an extra degree of branching in cassava starch or trace levels of contaminating activity in the enzymes capable of hydrolyzing glucosidic linkages other than α-D-(1→6).

The optimization range for starch solids concentration was selected to reflect starch gel concentrations reported in the literature during RS III synthesis. High starch concentrations are preferred because they facilitate drying and recovery of the product [6, 7, 9, 10]. *Shi* et al. [7] proposed starch gels containing 10–40% (w/v) starch solids, depending on the type of starch, for synthesis of a slowly digestible starch product. Native cassava starch granules, however, exhibit high moisture absorption and swelling during heat treatment, consequently forming highly viscous gels that are difficult to handle at solid concentrations above 20% (w/v). Furthermore, when the moisture content of the starch-water mixture is low, complete disruption of the starch granules is not possible and only partial gelatinization is achieved because the order-disorder transition characteristic of heat-gelatinization of starch is moisture dependent [26]. The degree of gelatinization of autoclaved starches containing starch solids concentration chosen for the optimization trials ranged between 97-99% as determined by the alkaline solubilization method [27] and decreased with increasing solids concentration beyond 20% (w/v).

Regression models fitted to explain the variability of DH of cassava starch were significant ($p<0.001$) for both enzymes. Polynomial functions of the form

$$DH = \beta_0 + \beta_1 C_1 + \beta_2 C_2 + \beta_3 C_3 + \beta_{11} C_1^2 + \beta_{22} C_2^2 + \beta_{33} C_3^2 + \beta_{12} C_1 C_2 + \beta_{13} C_1 C_3 + \beta_{23} C_2 C_3 + e$$

where C_1 is starch concentration, C_2 is enzyme concentration, C_3 is incubation time, β_{ij} is the regression coefficient, and e is the random error, were initially fitted. The coefficients of determination were 0.95 and 0.92 for isoamylase and pullulanase, respectively. For each data set insignificant ($p>0.05$) effects, mainly interactions, were stepwise removed from the model until a combination with only significant factors remained. Coefficients of determination of the adjusted models for isoamylase and pullulanase were 0.90 and 0.89, with respective standard errors of estimation of 0.15 and 0.22 and absolute errors of 0.16 and 0.17, respectively. The lack-of-fit errors were not significant ($p>0.05$) implying that adjusted models sufficiently represent the observed data. A comparison of observed and model DH values (Tab. 2) shows that relative deviations between fitted and observed values were generally below 5%, indicating that the fitted models are sufficient to predict DH at a 95% probability level. Tab. 3 summarizes the regression coefficients of the fitted models and their statistical significance. Enzyme concentration and incubation time exhibited significant linear and quadratic effects ($p<0.05$) and the DH increased with increasing enzyme concentration or incubation time but decreased with increasing starch concentration (Figs. 1 and 2).

Enzymatic hydrolysis of starch is influenced by the rate of enzyme diffusion, the accessibility of the substrate, substrate recognition leading to the formation of enzyme-substrate complexes and the proper catalytic action [28]. Whereas low starch solids suspensions mimic a nearly homogeneous phase hydrolysis, because both substrate and enzyme are almost in solution, high gel viscosity, created by gelatinizing cassava starch, limits enzyme dif-

Tab. 2. Observed and fitted values for degree of enzymatic hydrolysis of cassava starch.

Trial^x	Degree of hydrolysis [%]						Resistant starch content [g/100g]			
	Isoamylase			Pullulanase			Isoamylase		Pullulanase	
	Observed value	Fitted value	Relative deviation [%]	Observed value	Fitted value	Relative deviation [%]	Unpurified	Purified	Unpurified	Purified
1	4.93	4.75	-3.72	5.04	4.98	-1.22	50.56^{h*}	58.49^{f}	44.94^{g}	55.83^{f}
2	4.94	5.28	6.86	5.46	5.68	4.11	52.81^{i}	60.72^{h}	34.15^{d}	36.22^{b}
3	4.70	4.47	-4.98	4.73	4.70	-0.55	46.19^{ef}	50.00^{d}	35.45^{de}	51.80^{e}
4	5.18	5.00	-3.51	5.46	5.41	-0.92	52.85^{i}	56.93^{f}	44.26^{f}	54.94^{f}
5	4.00	3.65	-8.64	4.26	3.86	-9.43	36.94^{d}	35.11^{a}	30.10^{b}	44.06^{c}
6	4.42	4.19	-5.28	4.38	4.56	4.20	44.85^{e}	49.26^{d}	48.97^{h}	53.73^{ef}
7	3.33	3.37	1.32	3.61	3.58	-0.74	27.15^{b}	43.40^{c}	30.24^{ab}	52.12^{ef}
8	4.14	3.91	-5.65	4.22	4.29	1.64	34.83^{c}	43.21^{c}	37.12^{e}	53.74^{ef}
9	5.30	5.17	-2.46	4.98	5.02	0.74	53.44^{j}	64.10^{i}	49.52^{h}	62.36^{h}
10	4.72	4.70	-0.46	4.53	4.79	5.65	46.81^{f}	54.65^{e}	45.64^{g}	64.01^{h}
11	2.72	3.05	12.23	3.60	3.69	2.60	22.0^{a}	50.19^{d}	27.97^{a}	48.28^{d}
12	4.87	4.90	0.64	5.64	5.58	-1.10	57.26^{k}	59.26^{g}	35.77^{e}	61.35^{gh}
13	3.42	3.72	8.82	3.42	3.71	8.62	34.43^{c}	37.62^{b}	31.75^{c}	29.33^{a}
14	4.56	4.62	1.31	5.17	4.91	-5.11	48.65^{g}	59.08^{fg}	43.26^{f}	59.12^{g}
15-23	4.66-5.02^{y}	4.93	-1.71-5.88^{z}	4.65-5.41^{y}	5.02	-7.27-7.89^{z}	47.43^{fg**}	58.06^{f**}	43.39^{f**}	55.70^{f**}

^{x} Experimental conditions in Tab. 1; ^{y} Range of DH values at midpoint factor levels; ^{z} Range of relative deviation at midpoint trials *Values along the same column with similar superscripts are not significantly different at 95% confidence interval.

** Mean of RS III contents of mid point replicates.

Tab. 3. Regression coefficients of the fitted models for the degree of hydrolysis.

Coefficient	Isoamylase		Pullulanase	
	Estimate	*p*	Estimate	*p*
Constant	1.3008	0.0032	2.2856	<0.0001
C_1	-0.0374	0.0037	-0.0366	0.0488
C_2	0.0635	< 0.0001	0.0837	0.0002
C_3	0.1667	< 0.0001	0.1682	<0.0001
C_2^2	-0.0004	< 0.0001	-0.0009	0.0383
C_3^2	-0.0042	0.0002	-0.0039	0.0007

C_1 Starch concentration; C_2 Enzyme concentration; C_3 Incubation time.

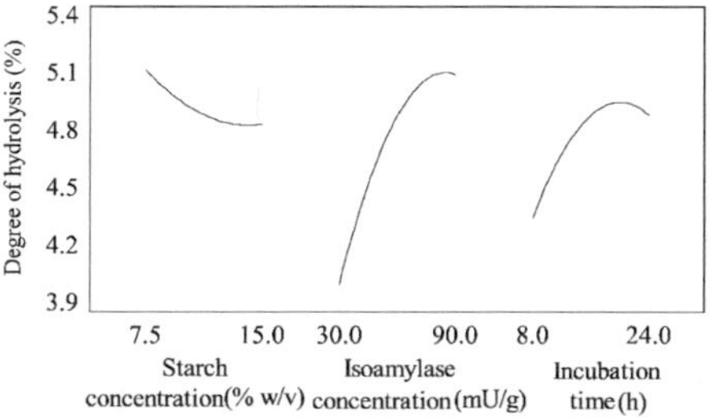

Fig. 1. Effects of starch concentration, isoamylase concentration and incubation time on hydrolysis of cassava starch.

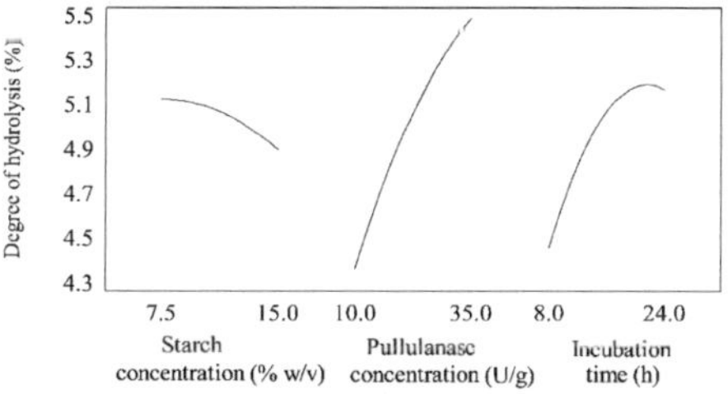

Fig. 2. Effects of starch concentration, pullulanase concentration and incubation time on hydrolysis of cassava starch.

fusion and thereby the ability of starch to bind on the active sites of the respective enzymes resulting in a decrease in DH. Consequently, high enzyme concentrations were required to increase the surface onto which the substrate could bind and guarantee catalytic action. Similarly, the long incubation times required to attain complete or near complete hydrolysis of the starch is an indicator for a slow rate of enzyme diffusion imposed by the highly viscous cassava gels. Whereas isoamylase is capable of hydrolyzing both the inner and outer α-D-(1→6) glucosidic linkages of amylopectin, the ability of pullulanase to cleave the inner longer α-D-(1→6) glucosidic linkages is limited particularly at low enzyme concentrations [18]. Cassava starch contains amylopectin, whose A- and B-α-D-(1→4) chains comprise 9-12 and 17-62 glucose units, respectively, and six to eight branch chains with an average chain length of 410-550 glucose units per molecule in the amylose polymer [13, 29]. Higher molecular weight polysaccharides are favoured substrates for isoamylase. The minimum substrate requirement is a maltose or, preferably, a larger A-chain and at least a maltotriosyl group in the B- or C- chain, although the low molecular size branched glucans are rather slowly debranched [18]. The higher enzyme activity levels of pullulanase compared to isoamylase, which were needed to cause comparable DH within the experimental domain of this study, can be explained by the fact that the rate of hydrolysis of amylopectin by pullulanase is slower and starts at the short outer chains. Furthermore, the DH as influenced by solids concentration decreased at a decreasing rate for isoamylase, but at an increasing rate for pullulanase implying that the two enzymes hydrolyzed cassava starch following different mechanisms.

Additional experiments, with conditions differing from those used to create the models, were carried out to evaluate the validity of the fits (Tab. 4). Starch concentration was maintained at 11.25% (w/v), because the corresponding regression coefficients were the least significant, whereas enzyme concentration and incubation time were selected from within and outside the model domain. Deviations between observed and fitted DH values for both enzymes were on average within the 5% confidence limit. This indicates that the models can be used to predict the hydrolysis of cassava starch at factor levels within and outside the experimental region and are, therefore, applicable in selection of conditions for the optimization of incubation time and enzyme availability within a wide range of factor levels. An optimum DH of 5.16 corresponded to 19.98 h incubation of the starch gel with 84.16

Tab. 4. Validation values of fitted models for determination of degree of hydrolysis of cassava starch.

Enzyme concentration		Starch concentration [%, w/v]	Time [h]	Degree of hydrolysis [%]		Relative deviation [%][a]
				Fitted value	Observed value	
Isoamylase [mU/g]	105	11.25	4	4.14	4.30	3.86
	75	11.25	12	4.90	4.94	0.82
	45	11.25	20	4.61	4.36	−5.42
	15	11.25	28	3.30	3.43	3.94
Pullulanase [U/g]	41.25	11.25	4	4.48	4.56	1.79
	28.75	11.25	12	5.07	4.73	−6.71
	16.25	11.25	20	4.86	5.21	7.20
	3.75	11.25	28	3.87	3.65	−5.68

[a]Deviation of observed from fitted value.

mU/g of isoamylase (Fig. 3). On the other hand, the maximum pullulanase concentration (35 U/g) and an incubation time of 21.5 h were necessary to achieve optimal DH of 5.61 (Fig. 4). Although the optimum DH of cassava starch hydrolyzed using pullulanase was higher than the optimal DH of cassava starch hydrolyzed using isoamylase, the tendency of DH to increase with further increase in pullulanase was nonetheless evident.

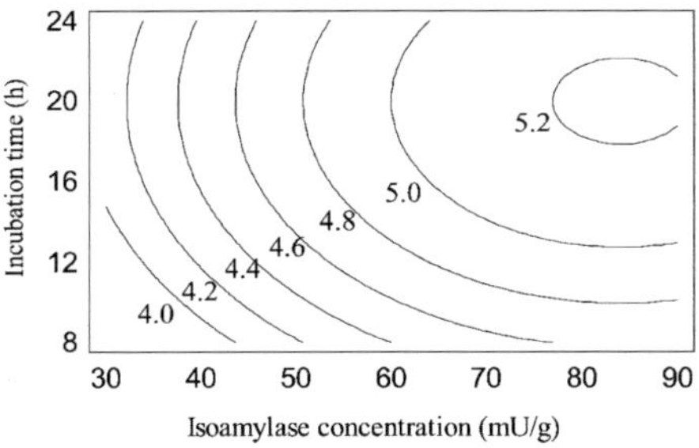

Fig. 3. Contours of estimated response surface for degree of hydrolysis at 11.25% (w/v) starch concentration using isoamylase.

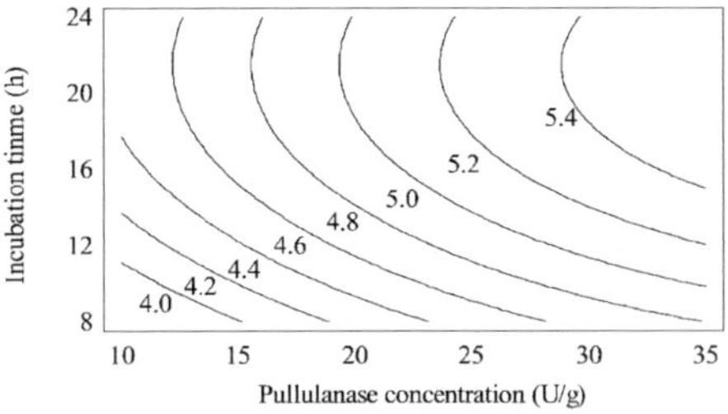

Fig. 4. Contours of estimated response surface for degree of hydrolysis at 11.25% (w/v) starch concentration using pullulanase.

3.2 Blue value, total soluble carbohydrates and β-hydrolysis limit

The blue value, measured as absorbance at 680 nm, was 0.34 for native cassava starch and the maximum absorbance, λ_{max}, of the native starch-iodine complex occurred at 565 nm. These values were lower than those observed by *Wickramasinghe* et al. [30], but within values reported by *Jyothi* et al. [31] and *Moorthy* [12]. The absorption intensity of the starch-iodine complex in native starches relates to the content and fine structure of amylose and amylopectin fractions and is primarily influenced by genotypic factors and conditions of field growth, harvesting and storage. High amylose content and the presence of long-chain branches in the amylopectin polymer are consistent with a high BV [11, 13, 23, 29]. Debranching of cassava starch significantly increased BV and λ_{max} to between 0.37–0.52 and 573-580 nm, respectively, depending on the DH, although the paired *t*-test did not show significant differences ($p>0.05$) between BV or λ_{max} of starches hydrolyzed using the two α-D-(1→6) glucanohydrolases. Following hydrolysis of α-D-(1→6) linkages, free linear α-D-(1→4) glucans are responsible for an enhanced iodine complexing capacity through the formation of double helices [32]. Debranched starches are, however, heterogeneous mixtures of glucans with different sizes, which comprise A, B and C amylopectin chains in addition to amylose strands. These absorb differently so that the resulting absorbance spectrum is a compilation of individual spectra, each of which show individual λ_{max} and absorbance values dependent on chain length. To obtain a representative estimate of the iodine-binding capacity of helical structures in the hydrolysates, starches were washed twice with deionized water, re-suspended, and the BV and λ_{max} was again analyzed. Blue value means increased to 0.60 and 0.57 in purified isoamylase and pullulanase hydrolysates, whereas λ_{max} increased by 21 and 10 nm, respectively, indicating that soluble carbohydrates comprising monomers, short chain oligosaccharides and suspended amylopectin depress the absolute absorbencies and λ_{max} of debranched starch. Furthermore, significant negative correlations between BV and total soluble carbohydrate content ($r = -0.69$, $p<0.001$; $r = -0.42$, $p<0.05$) were observed in both isoamylase and pullulanase debranched starches. Blue value and corresponding λ_{max} of purified starches were significantly higher in isoamylolysates than in pullulanase hydrolysates ($t = 2.11$, $p>0.05$; $t = 5.45$, $p<0.001$) and were significantly ($p<0.001$) positively correlated to DH in isoamylolysates but negatively correlated to DH in the pullulanase hydrolysates although the coefficient was not significant at the 95% probability level (Tab. 5).

The iodine-binding capacity of linear glucans depends on the degree of helix formation [32], whereas the wavelength of maximum absorbance is a function of the interaction of individual absorption spectra of the various polyiodide chains I^{3-}_{9}, I^{3-}_{11}, I^{3-}_{13}, I^{3-}_{15}, as they form complexes with linear glucans on the basis of their degree of polymerization [33]. A single iodine complex supporting helix comprises six glucose units per turn [34]. As the length of the glucan chain and hence the number of helical turns increases, the number of iodine molecules which can be accommodated also increases so that the iodine binding capacity is enlarged. Higher absorbance peaks in purified isoamylolysates imply a higher fraction of iodine accommodating double helices compared to pullulanase hydrolysates. Similarly, a higher λ_{max} is an indicator for the presence of longer helical structures in the isoamylolysates although an increase in iodine binding capacity also contributes synergistically to the shift of the λ_{max} to a higher level [33].

The β-amylolysis limit of native cassava starch was 60.2%. *Charoenkul* et al. [29] reported values of 75-80 and 57-63% for cassava amylose and amylopectin, respectively, which corresponded to the level of polymer branching, because the presence of α-D-(1→6) glucosidic linkages impedes complete conversion of starch into maltose by β-amylase. The β-amylolysis limit of isoamylolysates correlated positively with DH. The mean β-amylolysis limit values of hydrolysates having DH above 5.0 was 100.3±1.8% whereas the same value in pullulanase hydrolysates was 107.2±5.6%, probably due to concurrently occurring α-D-(1→4) hydrolysis. Furthermore, the correlations between DH and β-amylolysis limit, although significant ($p<0.05$) for both enzymes, indicate a weaker co-linearity for pullulanase debranched starches than for starches debranched with isoamylase (Tab. 5). It was therefore difficult to define the debranching end-point for cassava starch hydrolyzed by pullulanase from β-hydrolysis limit values. Following the conflicting observations regarding trends in BV, DH and β-amylolysis limit for pullulanase debranched starches, we sought to investigate the presence of contaminating activity in pullulanase using glycogen as substrate. An apparent activity of 5.26 U/mL at 50°C, pH 5 in the stock solution was assayed. Since pullulanase does not hydrolyze glycogen, it can be concluded that this activity is an indicator for the presence of enzymes capable of hydrolyzing α-D-(1→4) glucosidic bonds in the pullulanase solution. Our findings were in agreement with observations by *Bello-Pérez* et al. [35], who also reported partial debranching of amylopectin by pullulanase. Furthermore, although the amounts of degraded amylopectin increased with increasing enzyme concentration, the molecular size of pullulanase hydrolysates probably decreased with increasing enzyme concentration due to the contaminating activity.

Tab. 5. Correlation coefficients between degree of hydrolysis and covariate variables ($n = 23$).

	β-Amylolysis limit	Total soluble carbohydrate	Blue value		λ_{max}		Resistant starch type III	
			Before wash	After wash	Before wash	After wash	Before wash	After wash
Isoamylase								
R	0.88	-0.83	0.82	0.78	0.76	0.22	0.94	0.73
$P(R=0)$	<0.001	<0.001	<0.001	<0.001	<0.001	0.316	<0.001	0.001
Pullulanase								
R	0.68	0.40	-0.11	0.47	0.15	0.11	0.56	0.27
$P(R=0)$	<0.001	0.060	0.627	0.026	0.487	0.634	0.006	0.214

3.3 Resistant starch type III

Retrogradation of gelatinized starch results in a variety of ordered morphologies based on physical processes such as crystallization, vitrification, changes in molecular conformation, intermolecular association or combinations of each, involving multiple processes occurring in stages [36]. In contrast to granular starch, where crystallinity drops with increasing amylose content, recrystallization into enzyme-resistant morphologies is favoured in linear starches. The RS III contents of isoamylase and pullulanase debranched starches retrograded with and without prior purification are given in Tab. 2. The content of RS III in cassava starch isoamylolysates increased steadily with increasing hydrolysis of α-D-(1→6) linkages. An optimum RS III yield of 55.5% coincided with a DH of 4.96, which was closely comparable to the optimum DH of 5.16 in 11.25% (w/v) starch gels. The RS III yield was also significantly higher ($p<0.05$) in isoamylase debranched starches than in starches debranched using pullulanase.

A significant positive correlation exists between DH and RS III content but a higher coefficient of correlation was evident in isoamylolysates compared to pullulanase hydrolysates (Tab. 5). Furthermore, significant positive correlations between BV, λ_{max} and DH imply that the amounts of potential helix forming α-D-(1→4) polyglucans increase significantly with increased hydrolysis of cassava starch with isoamylase debranching at expense of the branched polymer. On the other hand, a negative correlation between DH and BV, and an insignificant change in λ_{max} of starches debranched with pullulanase imply possible interfering hydrolytic degradation of the debranched chains or impaired α-D-(1→6) hydrolysis of potential helix forming chains. Formation of RS III occurs when helical polymers aggregate to form double helices stabilized by hydrogen bonds [5, 34]. Whereas the realignment of linear glucans into helical polymers necessitates removal or segregation of the helix defects into regions of abundant chain defects, usually amorphous regions between lamellae, efficient aggregation of double helices into enzyme stable crystallites requires the removal of such defects from the crystalline regions to facilitate chain packing into crystalline order [34]. Amylopectin side branches are considered defects as they hinder proper alignment of the chains leading to ordering and crystallization. The proportional increase in RS III with increasing cleavage of α-D-(1→6) bonds observed for cassava starch isoamylolysates implies accelerated and more compact chain packing. The lower amounts of RS III in pullulanase hydrolyzed starches can be explained in terms of both chain length of helix forming glucans and extent of debranching. A mixture of both linear and branched residues results in an amorphous morphology as the branched molecules hinder spherulite formation and growth [36]. Furthermore, the presence of amorphous regions trapped within crystallizing helices interferes with hydrogen-hydrogen stabilization of aggregating strands. Similarly, short chains below the requisite length required for double helix formation do not recrystallize into stable RS III [14, 34]. Washing the debranched starches significantly increased RS III contents of hydrolysates (Tab. 2). This observation can be explained in terms of retrogradation kinetics, because purification removes buffer salts, monosaccharides and soluble oligosaccharides which impede retrogradation by inhibiting hydrogen bond formation between realigned chains or by acting as anti-plasticizers [14, 37, 38]. Consequently, the optimal chain length of 20 – 65 glucose units [9, 14, 15, 39] for effective retrogradation of starch gels can be approached by debranching, followed by purification through processes such as centrifugation, dialysis or filtration to remove reaction impurities and other low molecular weight by-products. Moreover, the RS III yield can be enhanced by optimizing the conditions under which the debranched material is allowed to retrograde.

4 Conclusions

Retrogradation of cassava starch into enzyme resistant structures after gelatinization of the granular starch is enhanced by α-D-(1→6) hydrolysis due to increased interaction between the linear starch polymers. The extent of enzyme-mediated removal of the branched polymer alongside the length of the resultant linear chains is, however, influenced by conditions of the catalytic transformation including the type of debranching enzyme, its purity and activity. The amount of RS III is further increased by removing low molecular weight by-products from the debranched starch which obstruct aggregation of enzyme resistant crystallites. Since a wide variety of linear α-D-(1→4) glucans of diverse chain length distribution form the RS III in debranched starches, further work should focus on characterization of the structural and physicochemical properties with a view to elucidating its techno-functional behaviour in end-use applications such as low energy food products and as prebiotics.

Acknowledgements

Christopher Mutungi would like to thank Deutscher Akademischer Austauschdienst (DAAD) for the financial support.

References

[1] C. C. Lim, L. R. Ferguson, G. W. Tannock: Dietary fibres as "prebiotics": implications for colorectal cancer. *Mol. Nutr. Food Res.* **2005**, *49*, 609–619.

[2] A. P. Nugent: Health properties of resistant starch. *Nutr. Bull.* **2005**, *30*, 27–54.

[3] D. L. Topping, P. Clifton: Short–chain fatty acids and human colonic function: roles of resistant starch and non-starch polysaccharides. *Physiol. Rev.* **2001**,*81*, 1031–1064.

[4] D. B. Thompson: Strategies for the manufacture of resistant starch. *Trends Food Sci.Technol.* **2000**, *11*, 245–253.

[5] S. G. Haralampu: Resistant starch - a review of the physical properties and biological impact of RS_3. *Carbohydr. Polym.* **2000**, *41*, 285–292.

[6] C. Onyango, C. Mutungi: Synthesis and in vitro digestion of resistant starch type III from enzymatically hydrolysed cassava starch. *Int. J. Food Sci. Technol.* **2008**, *43*, 1860–1865.

[7] Y- C. Shi, X. Cui, A. M. Birkett, M. G. Thatcher: U.S. Pat. 6929817B2 (**2005**), Slowly Digestible Starch Product.

[8] R. A. González-Soto, E. Agama-Acevedo, J. Solorza-Feria, R. Rendón-Villalobos, L. A. Bello-Pérez: Resistant starch made from banana starch by autoclaving and debranching. *Starch/Stärke* **2004**, *56*, 495–499.

[9] L. Haynes, N. Gimmler, J. P. Locke, M-Ra. Kweon, L. Slade, H. Levine: *U.S. Pat. 6613373B2* (**2003**), Enzyme Resistant Starch for Reduced-calorie Flour Replacer.

[10] B. W. Kettlitz, J. V. J.-M Coppin,. H. W. W. Röper, F. Bornet: *U.S. Pat. 6043229* (**2000**), Highly Fermentable Resistant Starch.

[11] A. L. Charles, Y. H. Chang, W. C. Ko, K. Sriroth, C. T. Huang: Influence of amylopectin structure and amylose content on the gelling properties of five cultivars of cassava starches. *J. Agric. Food Chem.* **2005**, *53*, 2717–2725.

[12] N. S. Moorthy: Physicochemical and functional properties of tropical tuber starches: a review. *Starch/Stärke* **2002**, *54*, 559–592.

[13] R. Hoover: Composition molecular structure and physicochemical properties of tuber and root starches: a review. *Carbohydr. Polym.* **2001**, *45*, 253–267.

[14] U. Lehmann, G. Jacobasch, D. Schmiedl: Characterization of resistant starch type III from banana (Musa acuminata). *J. Agric. Food Chem.* **2002**, *50*, 5236–5240.

[15] D. Schmiedl, M. Bäuerlein, H. Bengs, G. Jacobasch: Production of heat stable butyrogenic resistant starch. *Carbohydr. Polym.* **2000**, *43*, 183–193.

[16] R. C. Eerlingen, M. Deceuninck, J. A. Delcour: Enzyme-resistant starch. II. Influence of amylose chain length on resistant starch formation. *Cereal Chem.* **1993**, *70*, 345–450.

[17] H. M. Jespersen, E. A. Macgregor, B. Henrissat, M. R. Sierks, B. Svenson: Starch-and glycogen-debranching and branching enzymes. Prediction of the structural features of the catalytic $(\beta/\alpha)_8$-barrel domain and evolutionary relation to other amylolytic enzymes. *J. Protein Chem.* **1993**, *12*, 791–805.

[18] T. Yamamoto: *Enzyme Chemistry and Molecular Biology of Amylases and Related Enzymes* CRC Press, Boca Raton, Fl, **1994**.

[19] Y. Nakamura: Some properties of starch debranching enzymes and their possible role in amylopectin biosynthesis. *Plant Sci.* **1996**, *121*, 1–18.

[20] N. Nelson: A photometric adaptation of the Somogyi method for the determination of glucose. *J. Biol. Chem.* **1944**, *153*, 375–380.

[21] M. Somogyi: Notes on sugar determination. *J. Biol. Chem.* **1952**, *159*, 19–23.

[22] M. Dubois, A. K. Gilles, J. K. Hamilton, P. A. Rebers: Colorimetric method for determination of sugars and related substances. *Anal. Chem.* **1956**, *28*, 350–356.

[23] K. Thitipraphunkul, D. Uttapap, K. Piyachomkwan, Y. Takeda: A comparative study of edible canna (*Canna edulis*) starch from different cultivars. Part II. Molecular structure of amylose and amylopectin. *Carbohydr. Polym.* **2003**, *54*, 489–498.

[24] J. Tattiyakul, P. Pradipasena, S. Asavasaksakul: Taro *Colocasia esculenta* (L.) Schott amylopectin structure and its effect on starch functional properties. *Starch/Stärke* **2007**, *59*, 342–347.

[25] R. F. Tester, J. Karkalas, X. Qi: Starch - composition, fine structure and architecture. *J. Cereal Sci.* **2004**, *39*, 151–165.

[26] P. J. Jenkins A. M. Donald: Gelatinization of starch: A combined SAXS/WAXS/SANS study. *Carbohydr. Res.* **1998**, *308*, 133–147.

[27] A. Beleia, S. S. Butarelo, R. S. F. Silvia: Modelling of starch gelatinisation during cooking of cassava (*Manihot esculenta* Crantz). *Lebensm. Wiss. Technol.* **2006**, *39*, 399–404.

[28] P. Colonna, V. Leloup, A. Buleon: Limiting factors of starch hydrolysis. *Eur J. Clin. Nutr.* **1992**, *46*, S17–S32.

[29] N. Charoenkul, D. Uttapap, W. Pathipanawat, Y. Takeda: Molecular structure of starches from cassava varieties having different cooked textures. *Starch/Stärke* **2006**, *58*, 443–452.

[30] H. A. M. Wickramasinghe, S. Takigawa, C. Matsuura-Endo, H. Yamauchi, T. Noda: Comparative analysis of starch properties of different root and tuber crops. *Food Chem.* **2009**, *112*, 98–123.

[31] A. N. Jyothi, K. S. Kiran, B. Wilson, S. N. Moorthy, B. Nambisan: Wet Storage of Cassava Starch: Use of sodium metabisulphite and acetic acid and the effect on starch properties. *Starch/Stärke* **2007**, *59*, 141–148.

[32] C. A. Knutson: Evaluation of the variations in amylose–iodine absorbance spectra. *Carbohydr. Polym.* **1999**, *42*, 65–72.

[33] X. Yu. C. Houtman, R. H. Atalla: The complex of amylose and iodine. *Carbohydr. Res.* **1996**, *292*, 129–141.

[34] M. J. Gidley, D. Cooke, A. H. Darke, R. A. Hoffman, A. L. Russell, P. Greewell: Molecular order and enzyme-resistant retrograded starch. *Carbohydr. Polym.* **1995**, *28*, 23–31.

[35] L. A. Bello-Pérez, O. López-Paredes, P. Roger, P. Colonna: Molecular characterization of some amylopectins. *Cereal Chem.* **1996**,*73*, 12–17.

[36] J. A. Creek, G. R. Ziegler, J. Runt: Amylose crystallization from concentrated aqueous solution. *Biomacromolecules* **2006**, *7*, 761–770.

[37] R. C. Eerlingen, I. Van den Broeck, J. A. Delcour, H. Levine: Enzyme resistant starch. VI. Influence of sugars on resistant starch formation. *Cereal Chem.* **1994**, *71*, 472–467.

[38] K. Kohyama, K. Nishinari: Effect of soluble sugars on gelatinization and retrogradation of sweet potato starch. *J. Agric. Food Chem.* 1991, *39*, 1406–1410.

[39] R. C. Eerlingen, J. A. Delcour: Formation, analysis, structure and properties of type III enzyme resistant starch. *J. Cereal Sci.* **1995**, *22*, 129–138.

(Received: December 15, 2008)
(Revised: February 2, 2009)
(Accepted: February 4, 2009)

Appendix 3

Mutungi, C., Rost, F., Onyango, C., Jaros, D., & Rohm, H. (2009b)

Crystallinity, thermal and morphological characteristics of resistant starch type III produced by hydrothermal treatment of debranched cassava starch.

Starch-Starke, **61**, 634-645.

DOI 10.1002/star.200900167

Christopher Mutungi[a]
Felix Rost[a]
Calvin Onyango[b]
Doris Jaros[a]
Harald Rohm[a]

[a] Institute of Food Technology and Bioprocess Engineering, Technische Universität Dresden, Dresden, Germany
[b] Max Rubner Institute, Detmold, Germany

Crystallinity, Thermal and Morphological Characteristics of Resistant Starch Type III Produced by Hydrothermal Treatment of Debranched Cassava Starch

Cassava starch was debranched using pullulanase and the linear glucans recrystallized by incubation at 60°C or by temperature cycling at 120/60°C, and further subjected to heat-moisture treatment (HMT). Resistant starch (RS III) contents increased from 21.4 g/100 g in the debranched starch (DS) to 67.3 g/100 g in the debranched starch incubated at 60°C (DRS) and 47.8 g/100 g in the debranched starch subjected to temperature cycling (DCS), and further to 84.8 g/100 g and 88.4% g/100 g in HMT-DRS and HMT-DCS, respectively. Total crystallinity varied between 31.4-59.8% and the crystalline type was C in DS and DRS and A in DCS, HMT-DRS and HMT-DCS. The melting properties were characterized by broad endotherms, but the exact melting region and enthalpy were dependent on recrystallization method. The main endothermic peaks of DS and DRS occurred at 103.9 and 109.8°C, respectively, whereas DCS exhibited split endotherms at 113.6 and 138.1°C. Heat-moisture treatment broadened the endotherms and increased their enthalpies. Scanning electron micrographs revealed surface topography differences related to size and aggregation of individual crystalline bodies.

Keywords: Cassava starch; Hydrothermal treatment; Resistant starch type III

1 Introduction

Recrystallization of gelatinized starch into enzyme-resistant morphologies (RS III) is a result of the intimate packing of double helical starch into a partially crystalline system that resists the diffusion and binding of hydrolytic enzymes. This transition involves chain realignment and ordering processes whose kinetics exhibit strong dependence on time and temperature, and on the structural conformation of the starch polymers [1-3]. Consequently, a variety of ordered morphologies arise depending on the predominance of processes such as crystallization, vitrification, changes in molecular conformation, intermolecular association or combinations of these, involving multiple processes occurring in stages [4]. The structure of the re-ordered morphologies further influences the thermal stability and the physiological functionality of RS III [5-9]. Starch that resists digestion is of nutritional importance because of its reduced glycaemic and insulinemic responses as well as hypocholesterolemic and improved gut health effects. The latter arise primarily from the fermentation of RS III into short-chain fatty acids, biomass and other health-promoting metabolites by the lower gut microflora [10-12]. Whereas the molecular structure of the starch and therefore its recrystallization characteristics can be altered by pre- and/ or post-gelatinization modifications, retrogradation conditions influence the orientation and ordering of the starch polymers thus affecting the proportion of the starch that subsequently transforms into RS III, its physicochemical and morphological characteristics, and techno-functional properties [11, 13].

The digestibility of native starch depends on granular size, amylose /amylopectin ratio, amylopectin chain length, extent of intermolecular associations within the granule and the degree of crystallinity [14]. However, changes in susceptibility to hydrolysis by α-amylase occur when the starch is subjected to hydrothermal treatments due to alterations in granule stability, crystalline orientation and starch chain interactions. Based on moisture content, temperature and time, the hydrothermal treatment of starch can be categorized into gelatinization, annealing, heat-moisture treatment (HMT) and temperature cycling

Correspondence: *Christopher Mutungi*, Institute of Food Technology and Bioprocess Engineering, Technische Universität Dresden, 01062 Dresden, Germany. Phone: +49-351-463-36943, e-mail: chrismutungi@yahoo.co.uk.

[15-18]. During gelatinization, the starch containing excess water is heated above the gelatinization temperature to induce loss of crystalline order and the unfolding into a highly digestible polymer, whereas in annealing, the starch containing moisture above 65% (w/w) is maintained at a temperature between its gelatinization onset and glass transition temperatures (usually 40-55°C), a treatment that perfects the crystalline order and enhances resistance to enzyme hydrolysis. Under HMT, the starch is maintained above gelatinization temperature after moistening to a moisture content not exceeding 35% (w/w) so that modification is attained with minimum loss of granular integrity. Temperature cycling, on the other hand, subjects starch to an alternate heating and cooling process in which the time, the difference between the high and low holding temperatures and the number of heating and cooling cycles influence alteration of the starch properties [19, 20]. Previous investigations on effects of hydrothermal treatment on structural and physicochemical properties of starch have focused on native starches and few studies have been conducted on retrograded starch [15–17]. In this study, the effects of hydrothermal treatment on the enzyme resistance, thermal stability and morphology of debranched cassava starch were investigated with the aim of assessing its technological properties as potential functional RS III ingredient in foods.

2 Materials and Methods

2.1 Debranching and hydrothermal treatment of cassava starch

Cassava starch was obtained from Kenya Industrial Research and Development Institute, Nairobi, Kenya. The starch (10 g) was suspended in 70 mL 20 mmol/L sodium acetate-hydrochloric acid buffer, pH 5.0, and the resulting 12.5% (w/w) starch slurry gelatinized by autoclaving at 121°C for 15 min. The paste was cooled to 50°C and mixed with 10 mL pullulanase solution (Sigma P2986; 25 U/mL). The final starch and enzyme concentrations were 11% (w/w) and 25 U/g starch, respectively. The starch-enzyme mixture was incubated at 50°C for 24 h with stirring after which the debranched glucans were recovered by washing three times with excess cold deionized water accompanied by centrifugation at $1500 \times g$ for 10 min, freeze-dried and analysed for polymer chain distribution. The sample was labelled DS.

The optimum incubation temperature (60°C) for retrogradation of debranched cassava starch was determined in pre-trial experiments. The debranched cassava starch was weighed and added to deionized water to make 9, 17 and 23% (w/w) solids concentration. The slurries were autoclaved at 121°C for 15 min, then incubated at 60°C in a water bath for 2, 4, 8, 24 and 48 h after which they were freeze-dried. The samples were labelled DRS.

Temperature-cycling of debranched starch pastes (17%, w/w) was performed in three cycles at 121 or 135°C followed by incubation at 60°C to enhance nucleation and propagation of high temperature melting RS III. In the first cycle, starches were autoclaved at 121 or 135°C for 2 h and incubated for 22 h. In the second cycle samples were autoclaved for 2 h, incubated for 1 h, autoclaved again for 2 h and incubated for 19 h, whereas in the third cycle samples were autoclaved three consecutive times for 2 h, interrupted by 1 h of incubation after the first and second autoclave treatments and, finally, incubated for 16 h. The total handling time in each cycling treatment was 24 h. Samples were freeze-dried and labelled DCS.

Heat-moisture treatment was conducted on optimum RS III containing DRS and DCS. To determine the optimum HMT conditions, a $3 \times 2 \times 4$ factorial experiment in a completely randomized design was conducted using DRS. Freeze-dried starches (200 mg) were weighed into small screw-cap tubes and moistened to 10, 20 and 30% (w/w) by adding 22, 50 and 85 μL of deionized water, respectively. Samples were thoroughly mixed on a vortex mixer and allowed to equilibrate at 4°C for 16 h before autoclaving at 120 or 135°C for 0.5, 1, 2 and 4 h. Based on RS III contents, the optimum HMT conditions were selected and replicated on the DCS starch. The DS and optimally generated DRS, DCS, HMT-DRS and HMT-DCS were characterized using differential scanning calorimetry (DSC), wide angle X-ray diffraction (WAXS) and scanning electron microscopy (SEM). All measurements were made in duplicates.

2.2 Polymer chain distribution of debranched starch

Glucan chain length distribution of the debranched starch was determined using high-performance anion exchange chromatography equipped with pulsed amperometric detector, Bioscan 817. The starch (20 mg) was weighed into 2 mL vials and mixed with 400 μL of 2 mol/L aqueous sodium hydroxide solution while shaking vigorously on a vortex mixer to allow dispersion, before diluting with 1600 μL of double distilled water and maintaining the solution at 4°C and 450 rpm on a shaking Eppendorf® Thermomixer (Hamburg, Germany) for 24 h. An aliquot (1 ml) was further diluted with 9 mL of 200 mmol/L aqueous sodium hydroxide, filtered through a 0.45 μm filter and 1.0 mL auto-injected into a Metrosep A Supp 1 (4.6×250 mm) column (Metrohm, Filderstadt, Germany). The sample was eluted using an acetate gradient eluent comprising

100 mmol/L aqueous sodium hydroxide solution containing 10 mmol/L sodium acetate as eluent A, and 200 mmol/L sodium hydroxide solution containing 200 mmol/L sodium acetate as eluent B, at a flow rate was 1.0 mL/min. The elution gradient was 13% eluent B in the first 0-10 min, a linear incremental gradient of 13% to 100% eluent B between 10-65 min, and 100% eluent B between 65-115 min, a linear decremental gradient from 100% to 13% eluent B between 115 – 125 min, and finally 13% eluent B between 125 to 135 min. Kestopentaose was used as internal standard and areas under individual peaks integrated using Metrodata IC Net 2.3 software (Metrohm AG, Herisau, Switzerland).

2.3 Resistant starch content

Freeze-dried starch (50 mg) was weighed into pre-weighed 15 mL centrifuge tubes, suspended in 5 mL 0.1 mol/L sodium acetate-hydrochloric acid buffer, pH 5.2, containing 0.02% (w/v) sodium azide, vigorously homogenized and mixed with 200 μL enzyme solution containing 2400 U/mL of porcine pancreatic α-amylase and 60 U/mL amyloglucosidase. The enzyme solution was prepared by suspending 1 g porcine pancreatic α-amylase (Sigma A3176) in 0.1 mol/L sodium acetate buffer (pH 5.2) containing 0.02% (w/v) sodium azide and incubated at 37°C in a shaking water bath for 10 min. The suspension was centrifuged at 1500 × *g* for 10 min, the supernatant recovered and mixed with 2.5 mL amyloglucosidase (Sigma A7095) just before use [3]. The starch enzyme mixtures were incubated horizontally in a shaking water bath (37°C, 170 strokes/min) for 16 h and thereafter mixed with 5 mL of 96% (v/v) ethanol to denature the enzymes and precipitate the suspended starch. The mixture was homogenized and centrifuged at 1500 × *g* for 10 min and the pellet washed with 5 mL 80% (v/v) ethanol, freeze-dried and weighed after equilibrating to 25°C. The RS III content was determined as a percent ratio of the weight of the residue to the initial weight of sample.

2.4 Differential scanning calorimetry

Differential scanning calorimetry (DSC) measurements were done using DSC Q100 (TA Instruments Inc, Eschborn, Germany) calibrated with indium. Duplicate samples (3.0±0.1 mg) were weighed into hermetic aluminium pans, and 10 μL of deionized water was added. The pans were hermitically sealed and allowed to equilibrate overnight at ambient temperature. DSC runs were performed from 10 to 170°C at a heating rate of 10 K/min with an empty pan as reference. The data were analyzed using Universal Analysis 2000 V4.4 software (TA Instruments –Waters LLC).

2.5 Wide angle X-ray diffraction

Wide angle X-ray diffraction of freeze dried powders was performed using a Siemens D5000 (Siemens AG, Munich, Germany) X-ray diffractometer. Samples were scanned with Cu-K_α radiation (λ = 0.154 nm). The X-ray generator operating conditions were 40 kV and 25 mA. Samples were tightly packed in the sample holder and scanned in the angular range of 2θ = 2.5 - 70° with step intervals of 0.01° and a scanning rate of 6.5 s on each step. An amorphous sample, prepared by autoclaving a 10% (w/w) cassava starch-water suspension followed by freeze drying after storage at −82°C for 20 min was also scanned for estimation of the background scattering. Diffractograms were acquired using reflection geometry and peaks assigned and integrated using OriginPro 8 SR0 v8.0724 (OriginLab Corporation, Northampton, USA). Diffraction patterns were compared with patterns of standard allomorphs [21] and the areas of individual peaks integrated in diffractograms normalized between 2θ = 5 - 37.5°, whose crystalline phase had been acquired by subtracting the scattering of the normalized amorphous starch diffractogram (Fig. 1). The percent proportions of individual allomorphs were calculated as the total area under the allomorph specific peaks divided by the area of the sample diffractogram multiplied by 100, whereas the total percent crystallinity was the difference between the area under the sample diffractogram and the area under the amorphous starch diffractogram divided by the area under the sample diffractogram multiplied by 100.

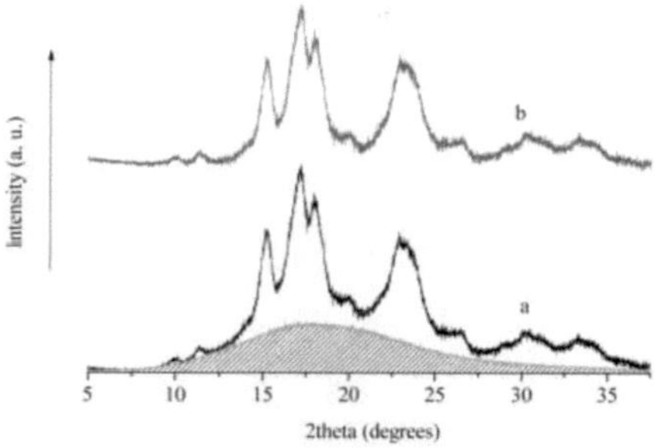

Fig. 1. Normalized X-ray diffractogram (a) fitted with amorphous scattering derived from gelatinized freeze dried cassava starch (b) after subtraction of the amorphous scattering.

2.6 Scanning electron microscopy (SEM)

Particles of freeze-dried powders were scanned using a DSM 982 Gemini scanning electron microscope (Zeiss,

Oberkochen, Germany). Samples were fixed with a brush to an electrically grounded conductive aluminium sample holder and covered with a layer of carbon. To avoid charges the sample obtained a sputter coating with a 5 nm thick layer of gold. Examination was done at an accelerating voltage of 10 kV over a working distance of 3-4 mm on an aperture of 30 nm.

2.7 Statistical analysis

Duplicate RS III values were subjected to multifactor ANOVA, whereas duplicate crystallinity and DSC parameter values were subjected to one-way ANOVA on GenStat 11th Edition ® statistical software (VSN International Ltd, Hemel Hempstead, UK) and means separated using Fisher's Least Significant Difference test at 95% confidence level.

3 Results and Discussion

3.1 Polymer chain distribution of debranched starch

Debranched cassava starch contained α-(1→4) glucans with DP of 5 – 48 (Fig. 2) with a high amount of chains of DP 10-24, amounting to 58.9% of the resolved chain lengths compared to 31.2%, the proportion of chains with DP >24. Polymer chains with DP <10 comprised only 8% due to purification of the starch after debranching. The mean chain length was 21.6±0.8 and corresponded well with the literature values for cassava amylopectin [22]. A base starch that comprises linear polymers of DP 10-35 is ideal for RS III formation [3].

3.2 Resistant starch content

Recrystallization of gelatinized starch into enzyme-resistant morphologies is influenced mainly by the molecular structure of the starch polymers, degree of branching, solids concentration and the recrystallization temperature and time [3, 17, 23-25]. The resistant starch content of gelatinized freeze-dried amorphous starch (AS) was 6.19 ± 0.39 g/100 g compared to 43.96 g/100 g in the native starch (NS) and 21.43 ± 2.72 g/100 g in DS. Whereas the resistant starch content in NS represents a slow glucosidic bond cleavage due to the molecular and physical structure of the starch granules [14, 17], the resistant starch contents in AS and DS represent re-associated polymers of the gelatinized starch. The higher amount of RS III in DS compared to AS implies that polymer association is enhanced by debranching. However as debranching proceeds, the long-chain amylose fraction participates in the recrystallization process by retrograding rapidly into ordered non-crystalline matrices that act as seed nuclei upon which multiple but separate short-chain crystalline domains are assembled [1]. The separate crystalline domains experience mobility constraints and a reduced ability to aggregate into perfect crystalline entities, hence, a partially cross-linked and loosely packed structure that exhibits limited resistance to enzyme hydrolysis is formed [24, 26]. The relatively low RS III content in DS indicates that, although substantial glucan interaction occurs during debranching, an incubation regime is necessary to allow for a more complete recrystallization.

The RS III contents of DRS are given in Tab. 1. In a pre-trial conducted to identify the optimum incubation temperature, RS III contents of starch (11%, w/w) debranched similarly with pullulanase (25 U/g starch) for 24 h

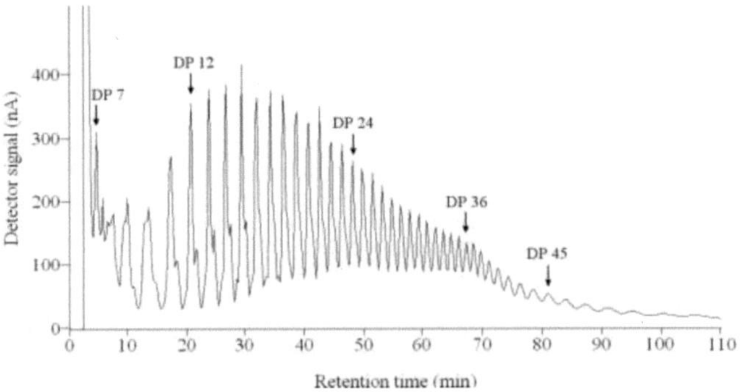

Fig. 2. α-Glucan chain length distribution of debranched cassava starch.

Tab. 1. Effect of starch solids concentration and retrogradation time on RS III content of debranched cassava starch (g/100 g).

Concentration [%, w/w]	Retrogradation time [h]				
	2	4	8	24	48
9	52.19[a*]	57.10[c]	58.98[d]	60.04[d]	61.59[e]
17	54.69[b]	58.59[d]	67.11[f]	66.73[f]	66.12[f]
23	54.49[b]	56.82[c]	60.63[de]	67.28[f]	66.13[f]

* Means with different superscripts are significantly different ($p<0.05$).

and thereafter incubated at 4, 30, 60 and 90°C were 23.15, 23.54, 28.59, and 25.25 g/100 g, respectively. Consequently, 60°C was selected as the retrogradation temperature. Contents of RS III in DRS were three times higher than the RS III content determined in DS and a significant interaction effect ($p<0.05$) between solids concentration and incubation time was observed. A maximum RS III content of 67% was achieved after incubation of starch pastes containing 17% solids for 8 h. This implies that during incubation, realignment and lateral packing of the glucans alongside other molecular interactions take place leading to completion of double helix formation, compartmentalization of the short linear glucans or enforcement of glucosidic linkages, hence, more stable and rigid aggregates [15]. Furthermore washing of the debranched starch at the recovery stage removes low molecular weight by-products and buffer salts whereas the brief autoclaving prior to incubation disrupts the previously imperfect crystallites thus facilitating a more effective re-association of the glucans [23]. The RS III content was higher than contents reported for banana, pea and rye starches, probably due to the shorter linear-chain glucans associated with these starches after debranching [23, 27, 28].

Resistant starch contents of debranched starch subjected to temperature-cycling increased significantly ($p<0.05$) with the number of autoclaving and incubation cycles. These were 43.1, 45.5 and 47.8 g/100 g in the 120/60°C cycling regime, and 38.4, 41.5 and 44.3 g/100 g in the 135/60°C regime for the first, second and third cycles, respectively. The RS III amounts, however, were lower than in DRS even after the third autoclaving and incubation cycle. Furthermore, RS III contents were significantly lower ($p<0.05$) in DS subjected to temperature cycling at 135/60°C compared to 120/60°C. This observation can be explained in terms of recrystallization kinetics. The high temperatures probably induce partial depolymerization of the glucans leading to substantial amounts of low molecular weight products that antagonize recrystallization or retard the nucleating ability of the glucans or sustain an exceedingly high molecular mobility that results in structures characterized by zones of weak crystalline order. *O'Brien* and *Wang* [29] observed that, although recrystallization of starch is enhanced during incubation at high temperature, such conditions allowed the development of surface fissures, which facilitate enzyme diffusion enabling both surface and internal erosion of the starch.

Recrystallization of gelatinized starch into RS III is a non-equilibrium thermo-reversible process governed by a three-phase mechanism of nucleation, propagation and maturation. Whereas the nucleation rate is increased by low temperatures close to the glass transition temperature and propagation/ maturation by the higher temperatures, recrystallization of linear glucans into stable RS III is enhanced when the retrogradation conditions do not stimulate nucleation and propagation of low melting amylose crystals [2]. Significant amounts of RS III are formed during propagation when excess free energy of the glucans generates the necessary internal forces for their relocation from the disordered to the ordered state and the subsequent rotation and arrangement of the glucans on the surface of the parent nuclei [1, 30]. The transition, however, relies on the formation of sub-critical nuclei of the crystalline phase by way of positive free energy of crystallization, a process dominated by conformational ordering and formation of double helices. The long linear glucans nucleate almost spontaneously after gelatinization without the need for low temperature incubation to induce the necessary change in the thermodynamic state [1]. Furthermore by hydrolyzing the α-D-(1→6) branches and using high starch solids concentration, lateral association and double helix formation is enhanced, thus enabling rapid attainment of the critical nuclei size at temperatures far above the glass transition temperature. Consequently, nucleation and propagation, which are both liquid-state events requiring orientational mobility of the glucans were optimized at 60°C and a solids content of 17-23% (w/w) provided a sufficient plasticization effect. Lower solids concentration (9%, w/w) yielded lower amounts of RS III probably due to excess plasticization that led to weak intra-helical linkages between coupling glucans.

The contents of RS III in HMT-DRS are presented in Tab. 2. The interaction effect between moisture content, temperature and treatment time was significant ($p<0.01$) and a significant increase in RS III contents occurred when the starch moisture was 20 or 30% (w/w) regardless of the temperature. A significant decline ($p<0.05$) occurred when heating exceeded 2 h at 135°C, probably due to formation of low molecular weight hydrolysis products or weak zones susceptible to hydrolysis [31]. The optimum

Tab. 2. Effect of moisture content, autoclaving temperature and time on RS III content (g/100 g) of heat-moisture treated debranched retrograded cassava starch.

Temperature [°C]	Moisture content [w/w]	Time [h]			
		0.5	1	2	4
121	10	39.58^{a*}	46.80^{b}	46.04^{b}	47.26^{b}
	20	65.56^{d}	71.87^{e}	83.65^{h}	81.07^{gh}
	30	75.28^{f}	75.14^{f}	84.81^{h}	82.89^{h}
135	10	39.38^{a}	46.56^{b}	53.63^{c}	54.53^{c}
	20	79.80^{g}	85.09^{h}	86.19^{h}	64.12^{d}
	30	82.75^{hi}	79.54^{g}	74.74^{f}	53.02^{c}

* Means with different superscripts are significantly different ($p<0.05$).

HMT conditions were a moisture content of 20% at 121°C and 2 h although no significant difference were observed between 20 and 30% moisture contents or 120 and 135°C at this HMT time level. A similar increase in contents of RS III, to about 80%, was reported following HMT of debranched banana, pea, Novelose 330 and rye starches [8, 23, 27, 28]. Lower moisture content, however, resulted in RS III contents that were lower than in the starting material (66 g/100 g) and only minimal increments were obtained with increased treatment time. *Chung* et al. [16] and *Gunaratne* and *Hoover* [17] have associated the increase in the susceptibility of some native starches to enzyme hydrolysis after HMT to the disruption of the weak double helices in the crystalline and amorphous regions. In line with this observation our results suggest that the enhancement of the crystalline order of linear glucans by HMT occurs in stages which probably include substantial unravelling of the weak helical glucans followed by their stabilization in the presence of a critical moisture content as opposed to native starches where such re-association may be hindered by the α-D-(1→ 6) branch points. Increase in enzyme resistance following HMT of the linear glucans can thus be attributed to enhanced interaction and the aggregation of helical structures by way of strong intermolecular forces forming a dense and compact supra-molecular structure which is unable to fit into the active site of α-amylase and one that has a reduced number adsorption sites for the enzyme. Moreover, besides the reduced accessibility of α-D-(1→4) glucosidic bonds, higher activation energies have to be overcome to achieve disentanglement of the structure during the enzyme adsorption stage.

3.3 Differential scanning calorimetry

DSC thermograms of native and debranched recrystallized cassava starches are shown in Fig. 3 and the transition parameters summarized in Tab. 3. The transition onset (T_o), peak (T_p) and conclusion temperature (T_c) and the transition enthalpy (ΔH) of NS corresponded well with literature values [14, 32]. Whereas T_o, T_p and T_c are a measure of crystalline stability and its variation within the starch granules, ΔH reflects the amount of double helical and crystalline order, a parameters that is influenced by amylose content, amylopectin chain-length distribution and amylose-lipid complexes [17]. Gelatinization of native

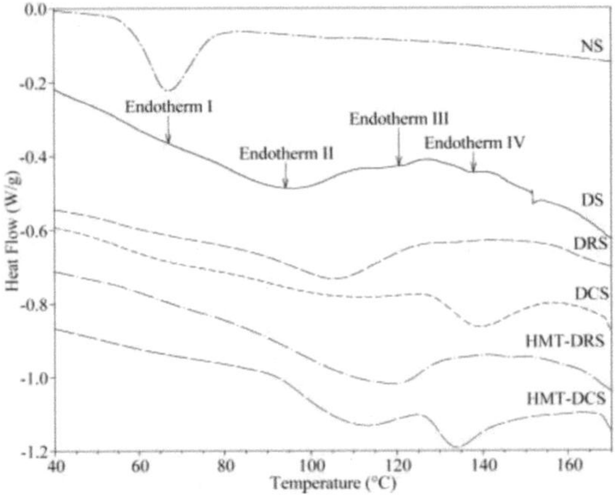

Fig. 3. DSC thermograms of native starch (NS), debranched starch (DS), debranched and retrograded starch (DRS), debranched and temperature-cycled starch (DCS), heat-moisture treated debranched and retrograded starch (HMT-DRS), and heat-moisture treated debranched and temperature-cycled starch (HMT-DCS).

Tab. 3. Thermal transition parameters of native, debranched, and hydrothermally treated debranched cassava starch.

Sample	Endotherm I				Endotherm II				Endotherm III				Endotherm IV				Total ΔH
	T_o	T_p	T_c	ΔH	T_o	T_p	T_c	ΔH	T_o	T_p	T_c	ΔH	T_o	T_p	T_c	ΔH	
NS	60.1	66.5	75.2	13.0	nd	nd	nd	nd	nd	nd	nd	nd	nd	nd	nd	nd	13.0
DS	52.1	63.1	70.7	0.8	80.8	103.9	115.7	8.5	114.7	119.7	123.1	0.4	132.1	136.8	139.8	0.3	10.0
DRS	49.3	59.7	73.2	1.3	85.6	109.8	121.7	12.7	nd	nd	nd	nd	127	136.7	141.5	0.6	14.9
DCS	51.4	64.3	75.0	1.3	89.9	113.6	123.5	3.7	130.2	138.1	154.5	6.6	nd	nd	nd	nd	11.6
HMT- DRS	60.1	69.2	75.1	0.1	83.0	117.3	137.5	23.2	nd	nd	nd	nd	nd	nd	nd	nd	23.3
HMT- DCS	51.3	62.1	78.5	0.4	94.6	112.4	125.2	11.6	130.6	137.6	158.5	7.0	nd	nd	nd	nd	19.0

T_o: transition onset temperature (°C), T_p: transition peak temperature (°C), T_c: transition conclusion temperature (°C), ΔH: transition enthalpy (J/g), nd: not detected.

starch is consistent with the disruption of double helical domains assisted by hydration and swelling of the amorphous regions [18]. Consequently, a single narrow endotherm is evident due to cooperative melting of crystallites within the starch granules. Contrastingly, debranched starch exhibited multiple endotherms with main endothermic peaks occurring at 103.9 and 109.8°C for DS and DRS, respectively (Fig. 3; Tab. 3). The endothermic multiplicity and the broad nature of the main transition endotherm (II) indicate crystalline inhomogeneity, consistent with crystallites of varying molecular stabilities. These could arise from the wide range of glucan chain lengths present in the starch (Fig. 2), and a probable disordered intermingling of the different size glucans creating crystallographic defects of varying intensities within individual crystallites. The endothermic peak shift from 103.9 in DS to 109.8°C in DRS, which was also accompanied by increase in ΔH from 8.5 to 12.7 J/g, supports the view that increased molecular ordering and glucan interactions occur during the incubation.

The endotherm I (50-80°C) can be associated with the melting of short-chain double helical structures which, in the debranched starch, lack the critical dimension necessary for inclusion into stable RS III [33] whereas endotherms III and IV at 119.7 and 136.8°C in DS and 136.7°C in DRS, probably, represent melting of RS III formed by the longer chain glucans. The low ΔH of the endotherms I, III and IV, however, indicate low double helical order, either because the chain length of crystallized glucans in these regions is shorter than that necessary for stable double helix formation or is longer than that required for formation of continuous double helical structures and/ or that the amounts of the respective chains in the starch are relatively low (Fig 2). A minimum glucan chain length of 10 glucose units is requisite for stable double helix formation whereas glucans longer than 100 glucose units form helical associations characterized by amorphous interruptions at the junction zones [3, 33].

Temperature cycling (DCS) resulted in two main transition endotherms (Fig. 3, Tab. 3). The emergence of an explicit transition endotherm III with T_p at 138.1°C, alongside the transition endotherm II with T_p at 113.6°C, suggested coexistence of two major fractions with different melting stabilities in the material. The decline in ΔH of the latter, however, is consistent with its heat destabilization, whereas the endotherm III at T_p 138.1°C is probably, in part, an enhancement of the minor endotherms III and IV registered for DS and DRS. The effect of temperature cycling can thus be described as one of facilitation of recrystallization of the amorphous long-chain fraction accompanied by the disruption of the unstable crystalline aggregates allowing formation, growth, perfection and completion of crystallization of more stable crystalline structures. This means that at temperatures similar to those used for debranching and retrogradation, the long-chain glucans, predominantly endogenous amylose, do not fully crystallize but rather exist as loose amorphous matrices caged within crystalline bodies of the shorter optimal chain glucans.

Heat-moisture treatment of recrystallized glucans resulted in increased ΔH and broadening of the transition endotherms (Fig. 3, Tab. 3). Of significance, however, was the decline in ΔH associated with the endotherm I in HMT starches, suggesting loss of its helical conformation or enhancement of its crystalline orientation and therefore incorporation into the main endotherm. The latter explanation is more plausible considering that the recrystallized HMT starches exhibited a broader transition and such stabilized short chain structures would represent, in part, the T_o of the new endotherm. HMT enhanced the endotherm II in DRS and DCS in which, ΔH increased two-fold from 12.7 to 23.2 J/g and three-fold from 3.7 to 11.6 J/g in HMT-DRS and HMT-DCS, respectively. Furthermore, T_p increased significantly ($p<0.05$) from 113.6°C in DRS to 117.3°C in HMT-DRS. Even so, such stability enhancement was not significant ($p>0.05$) for the endotherm III in DCS indicating that this fraction was characterized by high degree of crystalline order and that only the less stable helical or crystalline structures were sensitive to further stabilization by HMT.

Whereas the increase in ΔH can be explained in terms of increased ordering and stabilization of double helical structures through hydrogen bonds and other intermolecular forces [8, 34] the widening of the transition endotherm may be attributed to enhancement of crystallization that somehow, alters the architectural orientation of the crystallites [17]. Moreover, chain-length related modifications such as the lengthening of double helices as a result of twisting and completion of helix formation at the fringe ends of existing double helices, or adjoining of linear chains by way of extra linkages due to increased chain mobility within non-crystalline domains and/or perfections of the lattice imperfections result in alteration of the glass transition temperature and, hence, the melting parameters of the resultant RS III product [15]. However, the fact that the melting regions of recrystallized glucans remained relatively unchanged after HMT implies minimal internal relocation of the glucans probably because of the low starch moisture content which restricts molecular mobility. Accordingly, it can be argued that whereas the crystallite melting temperature increases with increased length of the glucans, the exact transition region and enthalpy are an interplay of unit chain-length distribution, the recrystallization temperature and the degree of crystallite perfection.

3.4 Wide angle X-ray diffraction

Fig. 4 shows the X-ray powder diffractograms of AS, NS and the debranched recrystallized starches. The respective diffraction peaks and percent crystallinities are summarized in Tab. 4. Native starch exhibited strong reflections at 2θ =15.18°, 17.24°, and 23.13° with a remotely resolved doublet at 18.17° and shoulder at 11.27° which typify A-type pattern. However, the intensity of the A polymorph signature peak at 2θ =18.17° was weak and no crystallinity could be apportioned to it. Furthermore, the reflections at 2θ = 30.84° and 33.61° implied presence of small regions of B-type packing, which amounted to 18.0% of the total crystallinity although the B-type signature reflection at $2\theta \approx$ 5.51° was conspicuously absent. The weak reflection at 2θ = 20.08° and absence of reflections at $2\theta \approx$ 7° and $2\theta \approx$ 13° implied indistinct amount of V-type crystallinity, usually associated with ordered single-helical glucans or linear glucans complexed with endogenous aliphatic lipids. Thus, NS was ascribed to the C_A crystalline type. Both A- and mixtures of A- and B-type packing patterns have been reported in native cassava starch [14, 32] due to variations in structure and the proportions of amylose/amylopectin usually determined by genotypic and field growth factors. The total crystallinity of 39.7% corresponded well with literature values of 38-40% [22], although lower values of 8–14% have also been reported, probably due to the method used to calculate the crystallinity [32].

The total crystallinity of DS was lower than that of NS (Tab. 4). The intensity of the reflection at 2θ = 15.18° degenerated to a shoulder (2θ=15.66°) whereas the magnitude of the peak at 2θ = 23.13° decreased and the peak was replaced by an unresolved doublet at 2θ = 22.35° and 23.37°. Furthermore, the peak intensity at 2θ = 17.24° significantly declined and the unresolved doublet at 2θ = 18.17° disappeared (Fig. 4), indicating that smaller and less perfect crystallites with a lower

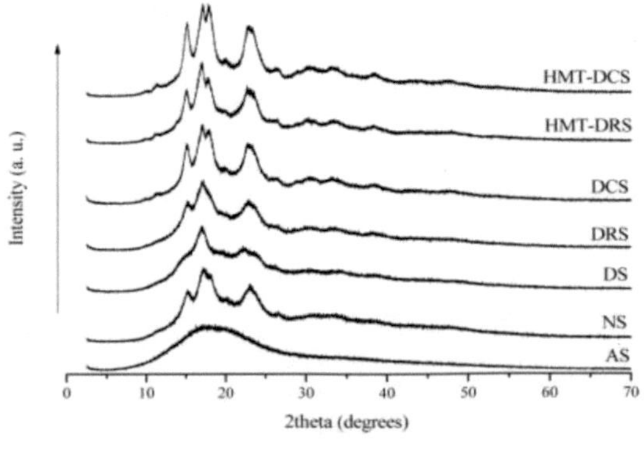

Fig. 4. X-ray diffraction patterns of amorphous starch (AS), native starch (NS), debranched starch (DS), debranched and retrograded starch (DRS), debranched and temperature-cycled starch (DCS), heat-moisture treated debranched and retrograded starch (HMT-DRS), and heat-moisture treated debranched and temperature-cycled starch (HMT-DCS).

Tab. 4. X-ray diffraction peaks and crystallinity of native, debranched, and hydrothermally treated debranched cassava starch.

Sample		Peaks and apportioned crystallinity [%] at diffraction angle [2θ]°											Allomorph type	Total crystallinity [%]
		10	11	15	17	18	20	22	23	26	30	33		
NS	P_d	nd*	11.27^{s}	15.18	17.24	18.17^{s}	20.08^{s}	nd	23.13	26.64	30.84	33.61	C_A	
	P_c	nd	0.83	5.56	12.74	nd	1.74	nd	10.25	1.42	3.55	3.60		39.69 ± 1.75
DS	P_d	nd	11.20^{s}	15.66^{s}	17.06	nd	19.67^{s}	22.35	23.37^{s}	26.13^{s}	30.69	33.68	C_B	
	P_c	nd	0.49	3.69	9.41	nd	1.20	4.36	3.97	2.57	3.15	2.58		31.42 ± 2.77
DRS	P_d	nd	11.31^{s}	15.32	17.13	nd	19.82^{s}	nd	22.95	26.42	30.48	33.64	C_A	
	P_c	nd	0.55	5.14	12.05	nd	1.39	nd	10.00	2.10	4.65	3.36		39.25 ±2.56
DCS	P_d	9.78	11.16	15.14	17.12	17.91	20.05^{s}	nd	22.81	26.32	30.49	33.22	A	
	P_c	0.26	0.68	7.02	9.62	8.08	2.00	nd	13.81	2.39	5.01	4.39		53.26 ±4.61
HMT-DRS	P_d	9.78	11.16	15.12	16.99	17.77	20.04	nd	22.67	26.51	30.48	33.22	A	
	P_c	0.35	0.67	7.35	10.03	7.62	1.73	nd	14.46	2.48	5.32	4.16		54.19 ± 4.15
HMT-DCS	P_d	9.95	11.30	15.13	17.12	17.91	20.04	nd	22.81	26.33	30.30	33.07	A	
	P_c	0.36	0.81	8.40	10.20	10.51	1.78	nd	15.76	2.07	5.56	4.33		59.80 ± 3.73

^{s} Peak not fully resolved but evident as a shoulder, nd: not detected, P_d: diffraction peak, P_c: crystallinity apportioned to peak.

ability to diffract X-rays compared to the native starch were dominant in DS [33]. However, the detection of a significant level of crystallinity in DS compared to AS is an indicator of occurrence of partial recrystallization during debranching. The DRS showed similar reflections as DS but a clear peak emerged at 2θ = 15.32° although its intensity was lower than that of NS. In addition, the reflection intensity at 2θ = 17.13° increased and the doublet at 2θ = 22.35° and 23.37° merged to a single peak at 2θ = 22.95°, whereas the intensities of reflections at 2θ = 30.48° and 33.64° increased although these remained relatively broad (Fig. 4). The reflection patterns of the DS and DRS thus denoted a C-type packing with the respective B-type crystallinities amounting to 44.7% and 20.4% of the total crystallinities and could therefore be distinguished as being C_B and C_A respectively, based on the relative contents of the real A- or B-allomorphs.

Contrastingly, DCS showed A-type crystalline packing characterized by a conspicuous doublet at 2θ = 17.12° and 17.91°. The reflection intensities were also higher resulting in a total crystallinity of 53.3%. This product, however, exhibited a higher digestibility than DRS, whose total crystallinity was significantly lower (39.3%). The C_B to A-type allomorphic transition and the increased enzyme susceptibility can be explained as having been induced by both temperature and dehydration effects. The lower melting B-type crystallites were unravelled and the loss of interhelical water in the central cavity was accompanied by rearrangement of double helices from hexagonal to monoclinic unit cell structure characteristic of the A-type allomorph. This double helical movement probably resulted in a microporous crystalline structure or crevices which would then allow better enzyme diffusion and increased binding sites [6, 17, 35].

Heat-moisture treated starches exhibited A-type crystallinity irrespective of the method of recrystallization and the total crystallinity was significantly higher in HMT-DCS than in HMT-DRS (Tab.4). Further to emergence of the doublet at $2\theta \approx$ 17° and 18°, the increase in intensity of the reflection at 2θ = 15.3° implied increased amount of A-type packing during HMT of both DRS and DCS. The increase in crystallinities coincided with higher amounts of RS III, which were 84.8 and 88.4% in HMT-DRS and HMT-DCS, respectively. Increase in crystallinity is consistent with development of crystalline domains that are sufficiently large and regularly arranged to diffract X-rays. The increase in individual reflections during HMT in the presence of low starch moisture content, however, implies that crystalline enhancement involves orientational modifications on the surface of crystallites rather than internal rearrangement.

Retrograded starch (RS III) crystallizes into A-, B-, C- or V-type polymorphs. High concentration, high temperature, short-chain glucans (DP <20) and presence of ethanol and lyotropic salts favour A-type crystallization, whereas low starch concentration, low temperature and the longer chain glucans (DP>20) promote B-type crystallite packing [5–7, 35-37]. Both A- and B-polymorphs are based on left-handed parallel stranded double helices arranged in

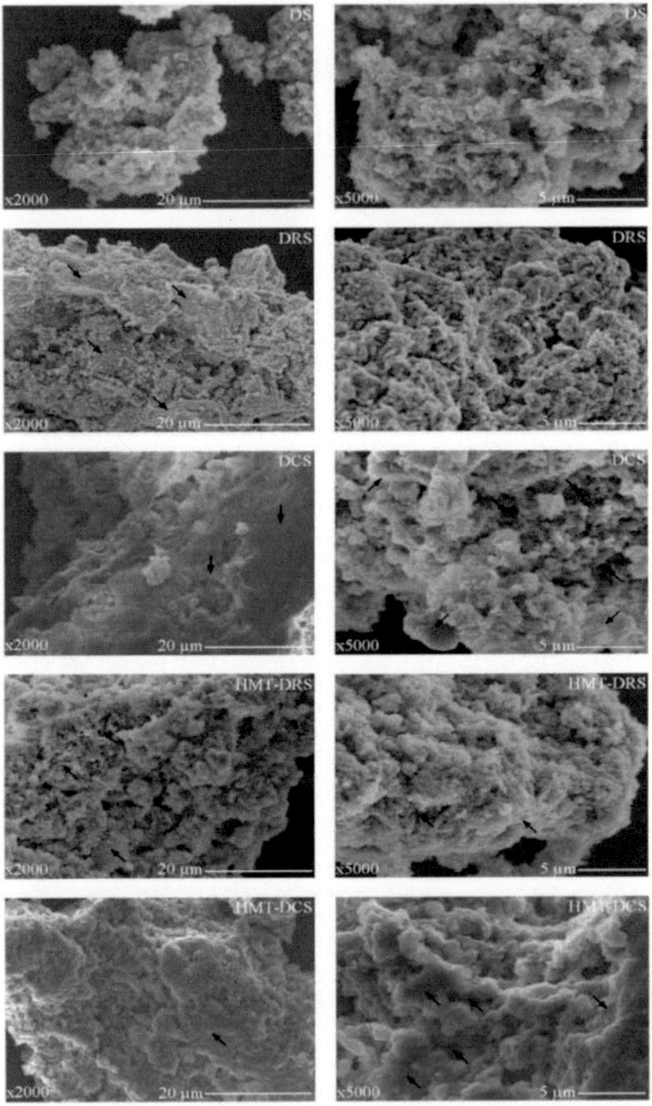

Fig. 5. Scanning electron micrographs of debranched starch (DS), debranched and retrograded starch (DRS), debranched and temperature-cycled starch (DCS), heat-moisture treated debranched and retrograded starch (HMT-DRS), and heat-moisture treated debranched and temperature-cycled starch (HMT-DCS). Arrows point at the distinct surface features.

monoclinic or hexagonal unit cells, respectively, and are further distinguished in that the A-type structure is more closely packed and less hydrated compared to the B-type packing in which a central void accommodates extra water molecules [38, 39]. Quantitative differences in the crystallinity of starches, however, can be occasioned by differences in crystal size, amount of crystalline regions, orientation of the double helical content within the crystalline domains, and extent of interaction between double helices or interplay of these factors. Furthermore, the total crystallinity encompasses intermediate crystalline structures, chain foldings, and lattice dislocations arising from the polydisperse nature of the glucans in the debranched starch which in turn have an impact on the thermal stability and enzyme susceptibility of the RS III [4, 24].

3.5 Scanning electron microscopy

Scanning electron micrographs of the recrystallized debranched starches are given in Fig. 5. The surface topography of DS comprises a fluffy structure characterized by small crystalline bodies that are loosely adhered to each other. This contrasts sharply with the surface profile of DRS in which the crystalline bodies are larger, more laterally ordered and closely packed. On the contrary, DCS exhibits larger crystalline bodies compared to DS and DRS and their aggregates possess both even and uneven surface zones induced by intense glucan mobility and rearrangement during temperature cycling. The even surfaces probably represent highly stable crystalline zones whereas the uneven regions typify surface porosity which is consistent with increased effective surface area for enzyme binding [6]. Following HMT, both DRS and DCS retained their differences in the size of the independent crystalline bodies, which, however, appear more closely adhered and possess a smooth surface lustre that is more pronounced in HMT-DCS compared to HMT-DRS. These SEM observations support the argument that HMT induces crystalline perfection by enhancing intermolecular forces rather than the physical relocation of the glucans.

4 Conclusion

Utilization of RS III in functional foods is governed by the ability to survive food processing conditions and to resist digestion in the upper gut upon ingestion. During enzymatic hydrolysis of starch, digestive enzymes need first to diffuse and bind to the solid substrate before bond cleavage. The structure of RS III dominates the reaction rate by affecting enzyme diffusion, binding and accessibility of the glucosidic linkages and is consistent with realignment, ordering and aggregation of linear glucans to form crystallites, whose type, orientation and perfection is dependent on the starch and retrogradation factors, and the post-retrogradation treatments. In this study, it was possible to produce RS III products having different heat stabilities and α-amylase susceptibilities. Although it is generally expected that A-type crystallites melt at higher temperatures than their B-type counterparts and that the B-allomorph is more enzyme-resistant than the A-allomorph, it is possible that other factors such as crystal size and extent of perfection, length of chains in the crystalline domains and the behaviour of the non-crystalline regions may have strong influence on the melting and digestibility patterns. However, the ability to control physicochemical properties of RS III by different recrystallization methods could be used to optimize RS III containing products. This knowledge can be used to produce a range of products with varying degree of resistance to α-amylase based on the recrystallization conditions. Further work should focus on investigation of the physiological and prebiotic properties of the hydrothermally treated cassava RS III with view to elucidating relationships between structure and physiological functionality.

Acknowledgements

Christopher Mutungi would like to thank Deutscher Akademischer Austauschdienst (DAAD) for the financial support.

References

[1] Y. Long, R. A. Shanks, H. Z. Stachurski: Kinetics of polymer crystallization. *Prog. Poly. Sci.* **1995**, *20*, 651–701.

[2] R.C. Eerlingen, M. Crombez, J. A. Delcour: Enzyme-resistant starch. I. Quantitative and qualitative influence of incubation time and temperature of autoclaved starch on resistant starch formation. *Cereal Chem.* **1993**, *70*, 339–344.

[3] D. Schmiedl, M. Bäuerlein, H. Bengs, G. Jacobasch: Production of heat stable butyrogenic resistant starch. *Carbohydr. Polym.* **2000**, *43*, 183–193.

[4] J. A. Creek, G. R. Ziegler, J. Runt: Amylose crystallization from concentrated aqueous solution. *Biomacromolecules* **2006**, *7*, 761–770.

[5] M. A. Whittam, T. R. Noel, S. G. Ring: Melting behaviour of A and B type crystalline starch. *Int. J. Biol. Macromol.* **1990**, *12*, 359–362.

[6] G. Williamson, N. J. Belshaw, D. J. Self, T. R. Noel, S. G. Ring, P. Cairns, V. J. Morris, S. A. Clark, M. L. Parker: Hydrolysis of A- and B- type crystalline polymorphs of starch by α- amylase and glucoamylase 1. *Carbohydr. Polym.* **1992**, *18*, 179–187.

[7] K. Shamai, H. Peled-Bianco, E. Shimoni: Polymorphism of resistant starch type III. *Carbohydr. Polym.* **2003**, *54*, 363–369.

[8] G. Jacobasch, G. Dongowski, D. Schmiedl, K. Müller-Schmehl: Hydrothermal treatment of Novolese 330 results in high yield of resistant starch type 3 with beneficial properties and decreased secondary bile formation in rats. *Brit. J. Nutr.* **2006**, *95*, 1063–1074.

[9] U. Lesmes, J. E. Beards, R. G. Gibson, M. K. Tuohy, E. Shimoni: Effects of resistant starch type III polymorphs on the human colon microbiota and short chain fatty acids in human gut models. *J. Agric. Food Chem.* **2008**, *56*, 5415–5421.

[10] C. C. Lim, R. L. Ferguson, G. W. Tannock: Dietary fibres as "prebiotics": Implications for colorectal cancer. *Mol. Nutr. Food Res.* **2005**, *49*, 609–619.

[11] M.G. Sajilata, R.S. Singhal, P. R. Kulkarni: Resistant starch-A review. *CRFSFS* **2006**, *5*, 1–17.

[12] A. H. Grabitske, J. L. Slavin: Gastrointestinal effects of low digestible carbohydrates. *Crit. Rev. Food Sci. Nutr.* **2009**, *49*, 327–360.

[13] D. B. Thompson: Strategies for the manufacture of resistant starch. *Trends Food Sci. Technol.* **2000**, *11*, 245–253.

[14] R. Hoover: Composition, molecular structure and physicochemical properties of tuber and root starches: A review. *Carbohydr. Polym.* **2001**, *45*, 253–267.

[15] L. Jayakody, R. Hoover: Effects of annealing on molecular structure and physicochemical properties of starches from different botanical origins - A review. *Carbohydr. Polym.* **2008**, *74*, 691–703.

[16] H-J. Chung, Q. Liu, R. Hoover: Impact of annealing and heat-moisture treatment on rapidly digestible, slowly digestible and resistant starch levels in native and gelatinized corn, pea and lentil starches. *Carbohydr. Polym.* **2009**, *75*, 436–447.

[17] A. Gunaratne, R. Hoover: Effect of heat-moisture treatment on the structure and physicochemical properties of tuber and root starches. *Carbohydr. Polym.* **2002**, *49*, 425–437.

[18] P. J. Jenkins, A. M. Donald: Gelatinization of starch: A combined SAXS/WAXS/SANS study. *Carbohydr. Res.* **1998**, *308*, 133–147.

[19] R. Wongsagonsup, S. Varavinit, J. N. BeMiller: Increasing slowly digestible starch content of normal and waxy maize starches and properties of starch products. *Cereal Chem.* **2008**, *85*, 738–745.

[20] A. Aparicio-Saguilán, E. Flores-Huicochea, J. Tovar, F. García-Suárez, F. Gutiérrez-Meraz, A. L. Bello-Pérez: Resistant starch-rich powders prepared by autoclaving of native and lintnerized banana starch: Partial characterization. *Starch/Stärke* **2005**, *57*, 405–412.

[21] A. Lopez-Rubio, M. B. Flanagan, P. E. Gilbert, J. M. Gidley: A novel approach for calculating crystallinity and its correlation with double helix content: A combined XRD and NMR study. *Biopolymers* **2008**, *89*, 671–768.

[22] R. F. Tester, J Karkalas, X Qi: Starch - Composition, fine structure and architecture. *J. Cereal Sci.* **2004**, *39*, 151–165.

[23] U. Lehmann, G. Jacobasch, D. Schmiedl: Characterization of resistant starch type III from banana (Musa ecuminata). *J. Agric. Food Chem.* **2002**, *50*, 5236–5240

[24] A. Pohu, V. Planchot, J.-L. Putaux, P. Colonna, A. Buléon: Split crystallization of maltodextrins at high concentration by isoamylase. *Biomacromolecules* **2004**, *5*, 1792–1798.

[25] A. González-Soto, R. Mora-Escobedo, H. Harnández-Sánchez, M. Sánchez-Rivera, L. A. Bello-Pérez: Influence of time and storage temperature on resistant starch formation from autoclaved debranched starch banana starch. *Food Res. Int.* **2007**, *40*, 304–310.

[26] M. Miao, B. Jiang, T. Zhang: Effect of pullulanase debranching and recrystallization on structure and digestibility of waxy maize starch. *Carbohydr. Polym.* **2009**, *76*, 214–221.

[27] U. Lehmann, C. Rössler, D. Schmiedl, G. Jacobasch: Production and physicochemical characterization of resistant starch type III derived from pea starch. *Nahrung/Food* **2003**, *47*, 60–63.

[28] C. Rössler, G. Jacobasch, D. Schmiedl, E. Gebhardt: Resistente Stärke Typ 3 aus Roggenstärke, ein funktionelles Lebensmittel. *Ernährung/Nutrition* **2002**, *26*, 297–305.

[29] S. O'Brien, Y.-J. Wang: Susceptibility of annealed starches to hydrolysis by α-amylase and glucoamylase. *Carbohydr. Polym.* **2008**, *72*, 597–607.

[30] J.-L. Putaux: Morphology and structure of crystalline polysaccharides: Some recent studies. *Macromol. Symp.* **2005**, *229*, 66–71.

[31] C. M. L. Franco, C. F. Ciacco, D. Q. Tavares: Effect of heat-moisture treatment on the enzymatic susceptibility of corn starch granules. *Starch/Stärke* **1995**, *47*, 223–228.

[32] S .N. Moorthy: Physico-chemical and functional properties of tropical tuber starches: A Review. *Starch/Stärke* **2002**, *54*, 559–592.

[33] M. J. Gidley, D. Cooke, A. H. Darke, R. A. Hoffmann, A. L. Russel, P. Greenwell: Molecular order and structure in enzyme-resistant retrograded starch. *Carbohydr. Polym* **1995**, *28*, 23–31.

[34] D. Cooke, M. J. Gidley: Loss of crystalline and molecular order during starch gelatinization: Origin of the enthalpic transition. *Carbohydr. Res.*, **1992**, *227*, 103–112.

[35] G. Zhang, M. Venkatachalam, B. R. Hamaker: Structural basis for the slow digestion property of native cereal starches. *Biomacromolecules* **2006**, *7*, 3259–3266.

[36] N. W. H. Cheetham, L. Tao: Variation in crystalline type with amylose content in maize starch granules: an X-ray powder diffraction study. *Carbohydr. Polym.* **1998**, *36*, 277–284.

[37] M. J. Gidley: Factors affecting the crystalline type (A-C) of native starches and model compounds: a rationalization of observed effects in terms of polymorphic structures. *Carbohydr. Res.* **1987**, *161*,301–304.

[38] Y. Takahashi, T. Kumano, S. Nishikawa: Crystal structure of B-amylose. *Macromolecules* **2004**, *37*, 6827–6832.

[39] D. Popov, A. Buléon, M. Burghammer, H. Chancy, N. Montesanti, J-L. Putaux, G. Potocki-Véronèse, C. Riekel: Crystal structure of A- amylose: A revisit from synchroton microdiffraction analysis of single crystals. *Macromolecules* **2009**, *42*, 1167–1174.

(Received: April 19, 2009)
(Revised: June 6, 2009)
(Accepted: June 12, 2009)

Appendix 4

Mutungi, C., Onyango, C., Rost, F., Doert, T., Jaros, D., & Rohm, H. (2010)

Structural and physicochemical properties and *in vitro* digestibility of recrystallised linear α-D-(1→ 4) glucans derived from mild-acid-modified cassava starch.

Food Research International, **43**, 1144-1154.

Food Research International 43 (2010) 1144–1154

Contents lists available at ScienceDirect

Food Research International

journal homepage: www.elsevier.com/locate/foodres

Structural and physicochemical properties and *in vitro* digestibility of recrystallized linear α-D-(1 → 4) glucans derived from mild-acid-modified cassava starch

C. Mutungi [a,*], C. Onyango [b], F. Rost [a], T. Doert [c], D. Jaros [a], H. Rohm [a]

[a] *Institute of Food Technology and Bioprocess Engineering, Technische Universitaet Dresden, D-01062 Dresden, Germany*
[b] *Max Rubner Institute, Schützenberg 12, D-32756, Detmold, Germany*
[c] *Department of Chemistry and Food Chemistry, Technische Universitaet Dresden, D-01062 Dresden, Germany*

ARTICLE INFO

Article history:
Received 1 November 2009
Accepted 7 February 2010

Keywords:
α-D-(1 → 4) glucans
Crystallinity
Thermal properties
Solubility
Water binding
Particle size
In vitro digestibility

ABSTRACT

Molecular structure and recrystallization method influence the techno-functional behaviour of recrystallized starch as a functional ingredient in foods. The physicochemical properties of debranched and recrystallized mild-acid-modified cassava starch were studied. Cassava starch was treated with 0.14 mol/L hydrochloric acid for 24, 96 and 216 h at 40 °C prior to debranching with pullulanase. The debranched starches (DS) were recrystallized by annealing (ANN-DS), temperature-cycling (TC-DS) or heat-moisture treatment (HMT-DS) and the particle distribution, crystallinity, thermal properties, solubility, water binding and *in vitro* digestibility were analyzed. Acid treatment increased the fraction of linear α-D-(1 → 4) glucans comprising 13–30 monomers. Particles comprised loosely to firmly coalesced primary elements forming aggregates of mono- or bi-modal size distribution at ⩽5 μm and ⩾20 μm. The relative crystallinities ranged between 31.1–56.1%. Water binding decreased significantly with acid treatment whereas both solubility and water binding were influenced by the recrystallization method and decreased in the order: DS > ANN-DS > TC-DS > HMT-DS. Major thermal transitions occurred at 80–130 °C and 130–160 °C, and the *in vitro* digestibility rates of 6.8–62.8% correlated significantly ($p < 0.001$) with relative crystallinity and melting enthalpy.

1. Introduction

The increasing use of recrystallized starch as a preferred source of dietary fibre in foods is based on its superior functional, textural and nutritional properties (Yue & Waring, 1998). Recrystallized starch, also referred to as enzyme-resistant starch type III, resists enzyme digestion in the upper gut and confers health benefits through moderation of metabolic (glycaemic, insulinemic, cholesterolemic) responses and improves digestive and colonic functions (Nugent, 2005). Foods enriched with RS III such as breakfast cereals and baked products (Baixauli, Salvador, Martinez-Cervera, & Friszman, 2008; Sanz, Salvador, Baixauli, & Fiszman, 2009), snacks, soups, sauces and confectionaries (Sajilata & Singhal, 2005), imitation cheeses (Arimi, Duggan, O'Riordan, O'Sullivan, & Lyng, 2008; Noronha, O'Riordan, & O'Sullivan, 2007), solid milk-based spreads (Ares, Baixauli, Sanz, Varela, & Salvador, 2009), and pasta (Gelenscér, Gál, Hódsági, & Salgó, 2008) are reported in the literature. In these products RS III is either used as bulking agent, fat replacer in low-fat or fat-free formulations, low-calorie replacer, prebiotic ingredient or viscosity and texture modifier.

RS III is formed when the linear starch polymers of gelatinized starch recrystallize by realigning into double helical strands that aggregate, and are stabilized by hydrogen bonds (Eerlingen & Delcour, 1995). Short linear α-D-(1 → 4) glucans of DP 20–35 are optimal for production of RS III with functional properties (Schmiedl, Bäuerlein, Bengs, & Jacobasch, 2000). *In vitro* synthesis of α-D-(1 → 4) linear glucans within the optimal range was demonstrated by Potocki-Veronese et al. (2005) and Schmiedl et al. (2000) using amylosucrase from *Neisseria polysaccharea* and sucrose as substrate; however no industrial production has been reported. Debranching of amylopectin rich starches is thus used as a more practicable alternative to achieving a near optimum glucan chain distribution for RS III formation (González-soto, Mora-Escobedo, Hernández-Sánchez, Sánchez-Rivera, & Bello Pérez, 2007; Lehmann, Jacobasch, & Schmiedl, 2002; Lehmann, Rössler, Schmiedl, & Jacobasch, 2003; Mutungi, Onyango, Jaros, Henle, & Rohm, 2009). Recovery and recrystallization of glucans was also shown to be enhanced by controlled treatment of native starch with mineral acids prior to debranching (Lehmann et al., 2003). However, the extent of acid hydrolysis of granular starches depends on the

* Corresponding author. Tel.: +49 351 463 36943; fax: +49 351 463 37126.
E-mail address: chrismutungi@yahoo.co.uk (C. Mutungi).

0963-9969/$ - see front matter © 2010 Elsevier Ltd. All rights reserved.
doi:10.1016/j.foodres.2010.02.008

acid and starch properties, and the reaction conditions (Jayakody & Hoover, 2002; Singh & Ali, 2000).

The structural characteristics of RS III are important because they influence the processing, sensory and functional behaviour of the products in which RS III is included. Processing phenomena such as pasting, melting, interaction with other macromolecular components and biological availability are also influenced by the qualitative and quantitative aspects of crystallinity of the starch (Colonna, Leloup, & Buleon, 1992). Thus, the aim of this work was to investigate the structure of RS III synthesized through a combination of mild-acid and enzymatic hydrolysis followed by recrystallization under different conditions. The structural and techno-functional properties as well as *in vitro* digestibility of the RS III were evaluated.

2. Materials and methods

2.1. Production of acid-modified cassava starch

Cassava starch was obtained from Kenya Industrial Research and Development Institute, Nairobi, Kenya. The extraction method and chemical composition of the starch are described in an earlier publication (Onyango & Mutungi, 2008). Starch (40 g) was suspended in 60 mL of 0.14 mol/L hydrochloric acid and incubated at 40 °C for 24, 96 and 216 h with occasional shaking every 10–12 h to re-suspend the starch. The starches were neutralized to pH 7 using 1 mol/L sodium hydroxide and centrifuged at 1500g. The pellets were washed thee times in excess deionized water and dried to constant weight in an air-oven maintained at 40 °C. The starches (10 mg) were mixed with 1 mL of 95% dimethylsulfoxide and dispersed by stirring (500 rpm) in an Eppendorf® Thermomixer (Eppendorf-Netheler-Hinz GmbH, Hamburg, Germany) at 95 °C for 2 h and further at 25 °C for 48 h. Absolute ethanol (9 mL) was added to precipitate the starches, and the mixtures were then centrifuged at 4000g for 10 min. The supernatants were decanted and the pellets dispersed in 10 mL deionized water containing 0.02% sodium azide by heating at 95 °C for 30 min. Aliquots (1 mL) were filtered through 0.45 μm filter and the molecular weights determined by asymmetric flow field flow fractionation (AFFFF). The AFFFF system was an Eclipse 3 (Wyatt Technology, Germany), equipped with an isocratic pump 1200 series (Agilent Technologies Inc., Santa Clara, CA, USA), a miniDAWN™ multi-angle laser-light scattering (MALLS) detector (Wyatt, Technologies Corp., Santa Barbara, CA, USA), and a refractive (RI) index detector (Knauer GmbH, Berlin, Germany). The asymmetrical channel structure comprised an aluminium cover plate with polycarbonate, 350 μm thick polyester spacer, and regenerated cellulose membrane with a molecular cut-off size of 10 kDA (Milipore, Billerica, MA, USA). Sample aliquots (50 μL) were injected at flow rate of 0.2 mL/min and focused at a flow rate of 3 mL/min. Mili-Q water containing 0.02% sodium azide was used as carrier and channel elution conducted at a flow rate of 1 mL/min. A specific refractive index increment rate of 0.151 mL/g was used (Roger & Colonna, 1992). The MALLS and RI profiles were processed using the ASTRA software (Wyatt, Technologies Corp., Santa Barbara, CA, USA).

2.2. Starch debranching and polymer chain distribution

Starch suspensions (12.5% w/w) were prepared in 20 mmol/L sodium acetate–hydrochloric acid buffer, pH 5.0, and gelatinized by autoclaving at 121 °C for 15 min. The pastes were cooled to 50 °C and mixed with pullulanase (Sigma P-2986, Sigma–Aldrich Chemie GmbH, Steinheim, Germany) to provide final starch and enzyme concentrations of 11% w/w and 25 U/g, respectively. The starch-enzyme mixtures were stirred at 50 °C for 24 h and washed thrice with excess cold deionized water by centrifugation at 3000 g for 10 min and freeze-dried. The debranched starches were milled using mortar and pestle and labelled DS.

Polymer chain length distributions of the DS were determined using high performance anion exchange chromatography with pulsed amperometric detector (HPAEC-PAD). The samples (20 mg) were weighed into 2 mL vials and 400 μL of 2 mol/L sodium hydroxide added while shaking vigorously on a vortex mixer. The mixture was diluted with 1600 μL double distilled water and maintained on a shaking Eppendorf® Thermomixer (4 °C, 800 rpm) for 24 h. Aliquots (1 mL) were further diluted with 9 mL 200 mmol/L sodium hydroxide and 20 μL auto-injected into a Metrosep A Supp 1 (4.6 × 250 mm) column (Metrohm, Filderstadt, Germany). Elution was done with an acetate gradient eluent comprising of 100 mmol/L sodium hydroxide, containing 10 mmol/L sodium acetate as eluent A and 200 mmol/L sodium hydroxide containing 200 mmol/L sodium acetate as eluent B, flowing at a rate of 1.0 mL/min. The elution gradient profile was as follows: 0–10 min: 13% eluent B, 10–65 min: linear increment from 13% to 100% eluent B, 65–115 min: 100% eluent B, 115–125 min: linear decrement from 100% to 13% eluent B, and 125–135 min: 13% eluent B. Kestopentaose was used as internal standard and the areas under individual peaks were integrated using Metrodata IC Net 2.3 software (Metrohm AG, Herisau, Switzerland).

2.3. Recrystallization of debranched starches

Portions of the DS were recrystallized by annealing (ANN-DS), temperature-cycling (TC-DS) or heat-moisture treatment (HMT-DS). The ANN-DS samples were prepared by autoclaving 20% w/w DS slurries at 121 °C for 15 min and then incubated at 60 °C for 24 h in a water bath. The TC-DS samples (20% w/w) were prepared by three autoclaving (121 °C for 2 h) and incubation cycles (60 °C for 1 h in the first two cycles and 60 °C for 16 h in the third cycle) in a water bath. The HMT-DS samples were prepared by moistening the freeze-dried DS to 20% w/w using deionized water and autoclaving at 120 °C for 2 h. After each treatment, the samples were freeze-dried and finely ground using a motor and pestle.

2.4. X-ray diffraction

Wide angle X-ray diffraction was performed using a D5000 X-ray diffractometer (Siemens, Karlsuhe, Germany). The X-ray generator operating conditions were 40 kV and 25 mA. Samples were tightly packed in the sample holder and scanned with Cu Kα radiation (λ = 0.154 nm) in the angular range of 2θ = 2.5–70° with step intervals of 0.01° and a scanning rate of 6.5 s on each step. An amorphous sample was scanned for estimation of background scattering. The amorphous sample was prepared by freeze-drying an autoclaved cassava starch–water suspension (10% w/w) which had been stored at −80 °C for 20 min. Diffractograms were normalized between 2θ = 5–37.5° and relative crystallinities were calculated according to Lopez-Rubio, Flanagan, Shrestha, Gidley, and Gilbert (2008a).

2.5. Differential scanning calorimetry

Differential scanning calorimetry (DSC) measurements were made using a DSC Q100 instrument (TA Instruments Inc., Eschborn, Germany). An indium standard was used to calibrate the equipment. Duplicate samples (3.0 ± 0.1 mg) were weighed into aluminium pans and 10 μL of deionized water was added. The pans were hermetically sealed and allowed to equilibrate overnight at ambient temperature. DSC runs were performed from 10 to 180 °C at a heating rate of 10 K/min with an empty pan as reference. The data were analyzed using Universal Analysis 2000 V4.4 software.

2.6. Solubility and water-binding index

Solubility and water-binding index were determined according to the method by Tattiyakul, Pradipasena, and Asavasaksakul (2007) with modification. Starch powders (50 mg) were weighed into pre-weighed 15 mL centrifuge tubes, mixed with 5 mL of deionized water, and placed on a shaking (500 rpm) Eppendorf® Themomixer maintained at 25, 35, 45, 55, 65, 75, 85 or 95 °C for 30 min. The slurries were centrifuged at 3000 g for 10 min and an aliquot of the supernatant analyzed for total soluble starch content using phenol sulphuric acid procedure (Dubois, Gilles, Hamilton, & Rebers, 1956). The supernatant was decanted and the pellet drained by touching with an adsorbent paper wick. The pellet was weighed and the water-binding index calculated as the weight of the drained pellet (mg) divided by the difference between dry sample weight (mg) and total soluble carbohydrate (mg glucose × 0.9).

2.7. Particle size distribution

A HELOS laser diffraction spectrophotometer (Symptec GmbH, Clausthal-Zellerfeld, Germany) was used to determine the particle size distribution. The sample (100 mg) was suspended in deionised water (100 mL, 20 °C) in a glass beaker and the suspension placed on an ultrasound bath (Elma Hans Schmidbauer GmbH & Co. KG, Singen, Germany) for 5 min to disperse loose agglomerates. Aliquots of the suspension were added drop-wise into the laser diffractometer dispersion cell containing deionised water as dispersant to give an optical density of 5%. Measurements were taken after 30 s of stirring. The Fourier lens focal length was 100 mm and the wavelength of the Helium–Neon laser was 632.8 nm. Data was acquired using manufacturer's software WINDOX 5 (Symptec GmbH, Germany).

2.8. In vitro digestibility

Starch powders (50 mg) were weighed into 15 mL centrifuge tubes and mixed with 2.5 mL sodium acetate–hydrochloric acid buffer (pH 5.2) containing 0.02% sodium azide and pre-heated to 37 °C. An α-amylase solution was prepared by suspending 1 g (24 U/mg) porcine pancreatic α-amylase (Sigma A-3176) in 10 mL buffer and incubating the mixture at 37 °C for 10 min before recovering the supernatant by centrifugation at 1500g. The α-amylase solution (1 mL) was mixed with 125 μL (300 U/mL) amyloglucosidase (Sigma A-7095) just before use, and 200 μL of the enzyme mixture added to the starch suspensions. The starch–enzyme mixtures were incubated horizontally in a shaking water bath (170 strokes/min) maintained at 37 °C. Samples (100 μL) were drawn before adding the enzyme solution and after 20, 60, 120 and 180 min, and added to 900 μL of 96% ethanol to denature the enzymes and precipitate polymeric starch. The samples were agitated on a vortex mixer, stored at 4 °C for 1 h and centrifuged at 5000g (4 °C) for 10 min. The supernatant was analyzed for reducing sugar content against glucose monohydrate standards using the Nelson–Somogyi alkaline copper reduction method (Nelson, 1944). First order kinetic equations, $C = C_\infty\ (1-e^{-kt})$, where C is reducing sugar content (mg glucose/100 g starch), C_∞ the equilibrium C (t = 180 min), k the hydrolysis rate constant and t time (min) were fitted (Göni, Garcia-Alonso, & Saura-Calixito, 1997). Gelatinised freeze-dried starch powder served as the control. Relative digestibility (d_r) was calculated as the area under the kinetic function of a sample divided by the area under the kinetic function of the control and multiplied by 100. Readily digestible starch (RDS), slowly digestible starch (SDS), and resistant starch (RS) fractions were determined from the glucose levels at t = 0 min (G0), t = 20 min (G20), t = 120 min (G120) and the dry initial weight of the sample (S) using the expressions: %RDS = (G20 − G0) × 0.9 × 100/S, %SDS = (G120−G20) × 0.9 × 100/S and %RS = (S − 0.9 × (G120 − G0)) × 100/S, where 0.9 is the stoichiometric constant for starch from glucose contents (Englyst, Kingman, & Cummings, 1992).

2.9. Statistical analysis

All measurements were performed in duplicates, results are given as means of the separate determinations. Relative crystallinity, solubility, water-binding index, particle size and differential scanning calorimetry values were subjected to two-way ANOVA. The k and C_∞ parameters in the *in vitro* digestibility experiment were subjected to one-way ANOVA, whereas RDS, SDS, RS, and R_d values were subjected to two-way ANOVA. All ANOVA tests were done on GenStat 11th Edition® statistical software (VSN International Ltd., Hemel Hempstead, UK) and the means separated using Fisher's Least Significant Difference test at 95% confidence level. Linear correlations were tested on Statgraphics Centurion XV® professional edition (StatPoint, Inc., Northern Virginia, Washington DC).

3. Results and discussion

3.1. Production of acid-modified cassava starch

Fig. 1 shows the radius of gyration (R_g) and molar mass distributions of the acid-modified starches together with the native non-acid treated control. R_g is related to the volume occupied by a molecule in solution and reflects chain length and branching patterns of the polymers (Yoo & Jane, 2002). From Fig. 1a, it is evident that more than 75% of the constituent polymers in the acid treated starches possess a R_g smaller than that of the native starch. The average R_g values decreased in the order: native > 24 h > 96 h > 216 h, indicating a steady decline in the point assemblage of the constituent polymers with increasing acid reaction time. However, the R_g of about 15% and 25% of polymers in starches treated with acid for 96 and 24 h, respectively, were larger than that of the native counterpart. The number average ($M_n = 1.5 \times 10^5$) and weight average ($M_w = 5.3 \times 10^5$) molar masses of the native starch corresponded well with literature values (Fig. 1b; Moorthy, 2002). The starches treated with acid for 24, 96 and 216 h had M_n values of 6.6×10^5, 4.2×10^5, 1.1×10^5 g/mol; and M_w values of 1.0×10^7, 3.4×10^6, and 2.4×10^5 g/mol, respectively. These values corresponded to polydispersities i.e. (M_w/M_n) of 15.15, 8.10, and 2.18, whereas native starch had a polydispersity of 3.53. Furthermore, a peculiar high molar mass fraction was evident in the 24 h acid-hydrolyzed starch (Fig. 1b, arrow). These observations suggest that polymer degradation is accompanied by reverse-type reactions in which the degraded polymers either retrograde, physically aggregate with adjacent starch molecules or form inter- or intra-molecular glucosidic bonds (Atichokudomchai, Chinachoti, & Varavinit, 2004; Belitz & Grosch, 1999).

Native starch exists as a semi-crystalline polymer (Tester, Karkalas, & Qi, 2004). During acid reaction, hydroxonium ions readily attack the oxygen atom of the glycosidic bonds in the amorphous sites because the dense helical packing of the crystalline domains resists acid percolation and hinders conformational change of the D-glucopyranosyl unit from chair to half-chair configuration essential for hydrolysis of the bond (Hoover, 2000). The larger molecular masses of 24 and 96 h acid-modified starches can thus be attributed to depolymerization of easily accessible amylose. Such a reaction would generate polymer fragments that attain the critical thermodynamic potential for aggregation. However, the decline in molecular size after 216 h of acid reaction implies that the aggregates are irregular and bound by weak forces, and therefore undergo further degradation. A slow erosion of the starch

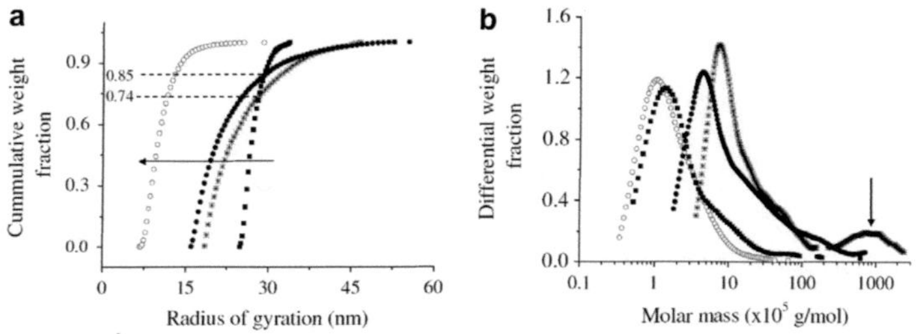

Fig. 1. Radius of gyration (a) and molar mass distribution (b) of native (■), and 24 h (∗), 96 h (●), and 216 h (○) acid-modified cassava starches. Arrows show decreasing radius of gyration (a) and a second molar mass distribution peak (b) due to aggregation. Broken lines indicate the cumulative weight fraction at the points where radius of gyration distributions intersect that of the native starch.

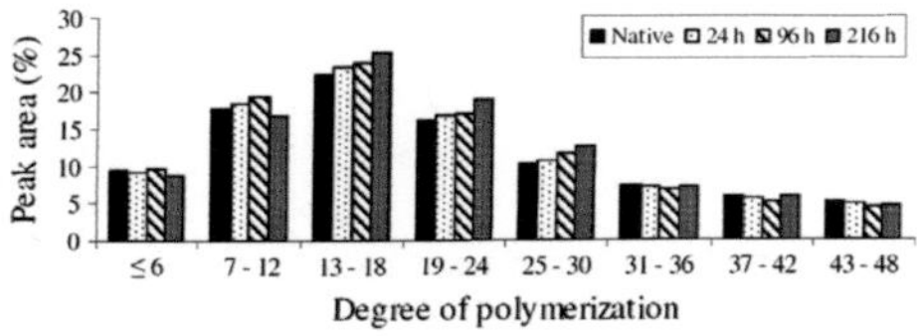

Fig. 2. Polymer chain distribution of native, and 24 h, 96 h, and 216 h acid-modified debranched cassava starches.

crystalline domains would also contribute to a decrease in molecular size but the tendency towards a normalized molar mass distribution with the peak around the same region as native starch indicates that these domains were not significantly affected.

The polymer chain length distribution profiles of the modified starches are shown Fig. 2. The relative proportion of glucans with DP 13-30 increased with increasing acid reaction time. This could be as a result of attainment of a more complete debranching of the acid depolymerized starches or acid fragmentation of the long amylose polymers. Lehmann et al. (2003) reported a similar change in the distribution of short linear polymers of acid depolymerized pea starch. The authors also observed a progressive decrease in the long chain amylose fraction although the detection limit of our HPEAC system could not allow quantification of this fraction in our samples. A linear glucan chain distribution of DP 10-35 is appropriate for butyrogenic RS III formation and a DP of 20-35 is reported to be ideal for enhanced thermal stability (Schmiedl et al., 2000).

3.2. X-ray diffraction

The wide angle X-ray diffractograms of the recrystallized glucan powders are shown in Fig. 3 and the relative crystallinities presented in Table 1. Regardless of acid modification of the granular starch, all DS exhibited C- type crystallinity as demonstrated by the peak at $2\theta = 17.2°$, the unresolved shoulder at $2\theta = 14.7°$ and the reflection doublet at $2\theta = 22.5°$ and 23.8°. This implies similar molecular orientation and structural packing of double helices. The peak at $2\theta = 17.2$ was, however, sharper and the intensity was significantly higher in the DS derived from 96 h acid-modified starch. The significantly higher relative crystallinity of this sample (Table 1) further suggests that larger amounts of regular crystalline material are present than in the other starches.

With regard to relative crystallinity, interaction effect between acid treatment time prior to debranching of the starch and the method of recrystallization was significant ($p < 0.01$). Annealing (ANN-DS) enhanced the shoulder at $2\theta = 14.7°$ to a full reflection peak at $2\theta = 15.2°$ whereas the doublet at $2\theta = 22.5°$ and 23.8° evolved to a single reflection at $2\theta = 23.1°$ (Fig. 3). This peak evolution pattern is an indicator of helical rearrangements resulting in larger amount of three dimensional ordered material and is supported by increases in relative crystallinity in comparison to DS (Table 1). Furthermore, the shift of the reflection to a higher Bragg angle means that there is change in the repeating unit i.e. the unit cell due to enhanced recrystallization, re-ordering or loss of moisture. Similar reflection patterns were also observed in TC-DS, but an extra reflection peak occurred at $2\theta = 18.0°$, indicating helical shift culminating in A-type packing. It is evident from the TC-DS diffractograms, however, that the intensity of this A-allomorph signature peak decreases with increasing acid reaction time, a trend that is also seen in the relative crystallinity values, where relative crystallinity significantly decreased with increasing acid treatment time (Table 1). This observation can be explained in terms of structural and thermodynamic influences. Han, BeMiller, Hamaker, and Lim (2003) reported the degradation of linear starch polymers following autoclaving with the longer chains being more vulnerable than the shorter chains which, on the contrary, exhibited aggregation tendencies. Excessively short chains, however, may demonstrate a low ability to nucleate and propagate into stable crystallites, particularly at temperatures that are too high above the glass transition temperature (Eerlingen, Crombez, & Delcour, 1993; Gidley & Bulpin, 1989).

All HMT-DS exhibited fully resolved A-type reflections at $2\theta = 9.9°$, 11.3°, 15.2°, 17.1°, 18.0°, 23.0°, 26.5°, 30.3° and 33.3°. Similar reflections for the A-type allomorph were reported (Lopez-Rubio et al., 2008a). The intensities of the reflections at $2\theta = 15.2$, 18 and 23° (Fig 3, HMT-DS) indicate an increasing degree of B–A allomorph conversion as the longer polymers of the granular starch become further depolymerized. The transition of B- to A-type packing during HMT is accompanied by loss of intra-helical moisture, thus enabling helical rearrangement without extensive molecular mobility due to the low moisture conditions (Le Bail, Bizot, & Buléon, 1993). This argument would suggest an inverse relationship between moisture binding capacity and development of A-type crystallinity which is consistent with the results on water binding indices of the starches (Fig. 4). Relative crystallinities of HMT-DS increased with increasing acid treatment time suggesting further that the removal of the long starch polymers through prior acid depolymerization enabled more effective crystallite enhancement by HMT (Table 1). The crystallinity enhancement pattern was, however, contrary to that observed for ANN-DS and TC-DS. This alludes to differences in the mechanisms for crystallite stabilization. The influence of recrystallization conditions on molecular

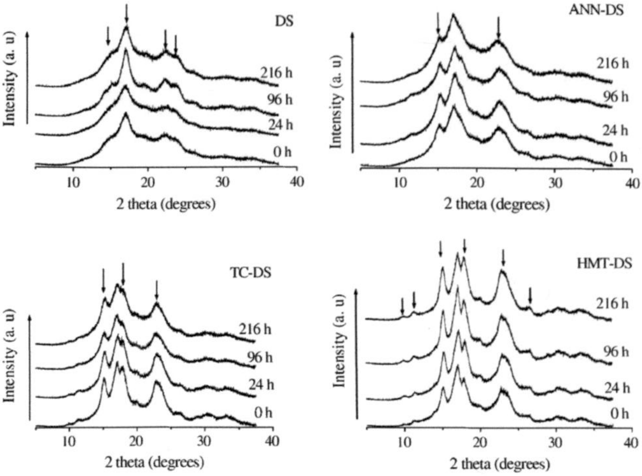

Fig. 3. X-ray diffraction patterns of debranched (DS), debranched and annealed (ANN-DS), debranched and temperature-cycled (TC-DS), and debranched and heat-moisture treated (HMT-DS) cassava starch powders. 0 h, 24 h, 96 h and 216 h represent acid treatment time of the granular starch before debranching. Arrows show evolution of reflection peaks.

Table 1
Relative crystallinities of debranched and recrystallized starch powders.

	Incubation time (h)			
	0[A]	24	96	216
DS	32.02^{a}_{v},[B]	31.12^{a}_{v}	38.34^{b}_{v}	30.49^{a}_{v}
ANN-DS	39.78^{a}_{x}	41.95^{a}_{x}	40.08^{a}_{v}	36.60^{a}_{x}
TC-DS	52.10^{a}_{y}	51.22^{a}_{y}	45.66^{b}_{x}	38.89^{c}_{x}
HMT-DS	51.88^{a}_{z}	54.87^{a}_{y}	54.79^{a}_{y}	56.14^{a}_{y}

DS: debranched, ANN-DS: debranched and annealed, TC-DS: debranched and temperature-cycled, HMT-DS: debranched and heat-moisture treated starch.
[A] Non-acid treated.
[B] Means in the same column sharing similar subscript letters or in the same row sharing similar superscript letters do not differ significantly ($p < 0.05$).

integrity and mobility of the helical glucans would probably impact on crystallite amount, type, size, orientation, and the presence and amount of crystallographic defects in the differently recrystallized starches, hence, the relative crystallinities as determined by the ability to diffract X-rays.

3.3. Differential scanning calorimetry

Table 2 summarises the thermal transition parameters of the glucan powders. A minor endotherm (not shown) occurred between 50–80 °C (ΔH = 0.8–1.5 J/g) in all the powders except HMT-DS. This endotherm is due to melting of short chain helical structures that are unable to form stable crystallites (Gidley et al., 1995). Major melting endotherms occurred in two temperature zones defined here as low temperature melting (LT, 80–130 °C) and high temperature melting (HT, 130–160 °C) regions for purposes of explaining the treatment effects. Melting in the HT region is associated with crystalline structures formed by the long amylose chains whereas melting in the LT region could arise from the shorter chain crystalline structures (Gidley et al., 1995) or imperfectly crystalline components of amorphous chains caged within relatively crystalline domains. The DS melted predominantly in the LT zone and only minor endotherms occurred in the HT zone. The total enthalpies, however, increased steadily with increasing acid modification of the granular starch up to 96 h. This can be associated with formation of relatively stable crystallites subsequent to depolymerization of the long chain polymers.

Similarly, the ANN-DS melted predominantly in the LT zone although the transition enthalpies increased, and the endotherms markedly shifted to the right, suggesting enhanced ordering of helical structures. The lack of a significant increase in the HT melting enthalpies indicates that the long chain fraction requires higher energies for mobilization into crystalline entities. This argument is supported by the thermal transition behaviour of TC-DS powders. However, whereas the HT endotherms were enhanced by TC, the LT enthalpies of TC-DS remained lower than the LT enthalpies of ANN-DS. This pattern is consistent with the melting of less stable crystallites and their reorganization into more stable fractions, probably on the basis of polymer chain lengths among other factors. Furthermore, the HT enthalpies decreased significantly with increasing acid depolymerization time, in line with the declining content of long range helical order. On the other hand, the fact that higher T_p and T_c in the HT endotherm occurred after some depolymerization of starch was achieved (Table 2) indicates that a certain polymer chain length is desirable for the formation of thermally stable crystallites. Moreover, double helices are formed within limited lengths of the linear polymers and large non-helical junctions between aggregating double helical strands would induce early transition and low enthalpies due to co-operative melting. HMT-DS exhibited broader melting temperature range characterized by shift of T_p and T_c in the LT endotherm to higher temperature. The endotherm broadening was accompanied by increased transition enthalpies (Table 2). This observation implies that higher structural stabilization of the linear glucans was induced by HMT as compared to ANN and TC. Contrastingly, no HT endotherms were formed in the HMT-DS, which suggests that the enhancement

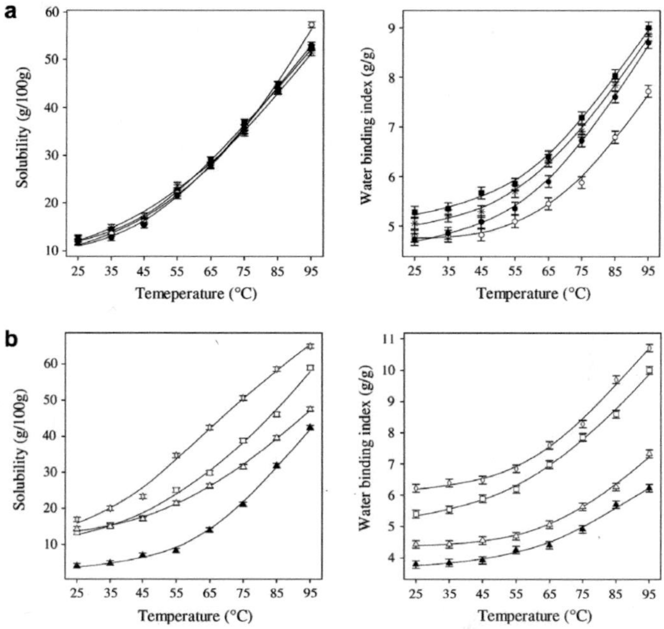

Fig. 4. Solubility and water binding profiles of debranched and recrystallized cassava powders (a) Effect of acid depolymerization: native (■), and 24 h (∗), 96 h (●), 216 h (○) acid-modified; (b) effect of recrystallization method. DS (◇), ANN-DS (□), TC-DS (△), HMT-DS (▲).

Table 2
Thermal transitions of debranched and recrystallized starch powders.

	DS				ANN-DS				TC-DS				HMT-DS			
	0 h^{A}	24 h	96 h	216 h	0 h	24 h	96 h	216 h	0 h	24 h	96 h	216 h	0 h	24 h	96 h	216 h
LT endotherm																
T_o	80.8^{a,B}	79.6^{a}	79.9^{a}	80.7^{a}	85.6^{b}	87.5^{b}	80.6^{a}	83.1^{a}	83.1^{a}	81.5^{a}	82.6^{a}	79.3^{a}	83.0^{a}	82.8^{a}	82.8^{a}	81.2^{a}
T_p	103.9^{b}	105.0^{b}	98.1^{a}	97.2^{a}	105.9^{b}	107.6^{c}	108.7^{c}	112.2^{d}	107.8^{c}	105.8^{b}	106.4^{b}	105.3^{b}	117.2^{e}	120.5^{e}	119.0^{e}	112.8^{d}
T_c	111.1^{a}	117.2^{b}	110.5^{a}	112.8^{a}	119.8^{b}	119.8^{b}	119.6^{b}	130.1^{e}	127.9^{d}	123.2^{c}	122.4^{c}	124.2^{c}	133.2^{e}	136.6^{f}	135.2^{f}	128.9^{d}
ΔH	7.1^{b}	8.9^{c}	8.4^{c}	5.6^{a}	12.7^{g}	11.9^{f}	13.5^{h}	12.9^{g}	6.9^{b}	8.9^{c}	9.6^{d}	10.6^{e}	24.3^{k}	27.1^{l}	23.5^{j}	22.2^{i}
HT endotherm																
T_o	132.1^{b}	133.7^{b}	133.2^{b}	130.8^{a}	150.5^{d}	138.3^{c}	138.1^{c}	137.8^{c}	129.7^{a}	129.4^{a}	132.9^{b}	128.5^{a}	nd	nd	nd	nd
T_p	136.8^{b}	144.2^{c}	145.5^{c}	133.5^{a}	153.7^{e}	146.6^{d}	147.8^{d}	147.0^{d}	137.9^{b}	142.2^{c}	146.1^{d}	135.4^{b}	nd	nd	nd	nd
T_c	141.3^{b}	161.2^{f}	158.9^{f}	136.4^{a}	157.1^{f}	160.3^{f}	159.5^{f}	154.8^{e}	150.7^{d}	156.1^{e}	155.3^{e}	146.5^{c}	nd	nd	nd	nd
ΔH	0.7^{a}	1.7^{b}	2.5^{c}	2.2^{c}	0.6^{a}	1.3^{b}	1.8^{b}	0.5^{a}	6.9^{e}	9.8^{f}	5.4^{d}	1.3^{b}	nd	nd	nd	nd
Total ΔH^{C}	8.6^{a}	11.6^{c}	12.1^{c}	9.6^{b}	14.9^{e}	14.3^{e}	15.1^{f}	14.4^{e}	13.1^{d}	19.5^{h}	17.0^{g}	14.8^{e}	24.3^{k}	27.1^{l}	23.5^{j}	22.2^{i}

DS: debranched, ANN-DS: debranched and annealed, TC-DS: debranched and temperature-cycled, HMT-DS: debranched and heat-moisture treated starch.
LT: low temperature, HT: high temperature, T_o: transition onset temperature (°C), T_p: transition peak temperature (°C), T_c: transition conclusion temperature (°C), ΔH: transition enthalpy (J/g), nd = not detected.

[A] Acid treatment time of the granular starch; 0 h represents the untreated starch.

[B] Means in the same row sharing the same superscript letters do not differ significantly ($p < 0.05$).

[C] Total ΔH: Total enthalpies, which includes enthalpy of a minor endotherm recorded at 50–75 °C, LT endotherm and HT endotherm.

of crystalline structure by HMT does not involve obvious helical mobility.

3.4. Solubility and water-binding index

The solubility and water binding profiles of the recrystallized glucans generally increased with elevated temperature. Fig. 4a compares solubility and water binding of the recrystallized glucans as influenced by acid reaction time. Solubility did not differ between 25 and 85 °C but acid modification for 216 h resulted in a significantly higher average solubility at 95 °C, which is consistent with a relatively higher proportion of short chain glucans. Water binding indices decreased significantly ($p < 0.05$) with prolongation of acid treatment time. Gidley and Bulpin (1989)

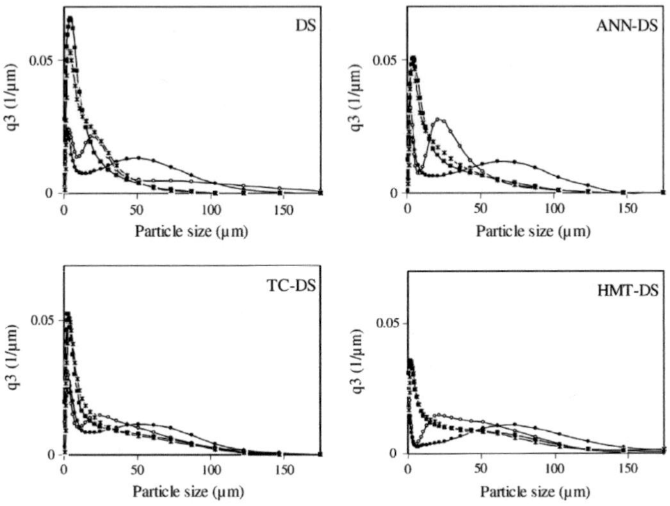

Fig. 5. Particle size distributions of native (■), and 24 h (*), 96 h (●) and 216 h (○) acid-modified debranched (DS), debranched and annealed (ANN-DS), debranched and temperature-cycled (TC-DS) and debranched and heat-moisture treated (HMT-DS) cassava starch powders.

demonstrated that long amylose chains form cross-linked networks whereas short polymers precipitate into dense crystalline entities as a consequence of extensive aggregation. The reaction of starch with acid selectively depolymerized the long starch polymers and enhanced aggregate formation as opposed to cross-linked matrices in which more water would be bound. A similar inverse relationship between water binding and molecular size of debranched starch was reported by Guraya, James, and Champagne (2001).

The effect of recrystallization method on solubility and water binding is given in Fig 4b. Both solubility and water binding decreased significantly ($p < 0.05$) in the order: DS > ANN-DS > TC-DS > HMT-DS. These differences can be attributed to differences in degree of associative bonding brought about by the different recrystallization conditions. The decrease in solubility and water binding of retrograded, heat-moisture treated or annealed granular and processed starches are reported in the literature (Adebowale, Henle, Shwarzenbolz, & Doert, 2009; Aparicio-Saguilán, Flores-Huicochea, Tovar García-Suárez, Gutiérrez-Meraz, & Bello-Pérez, 2005; González-Soto et al., 2007; Gunaratne & Hoover, 2002; Jayakody & Hoover, 2008), although, in some native starches, an increase in these parameters is also described (Jayakody & Hoover, 2008). Nevertheless, solubility and water binding are indicators of the extent of interaction between the starch polymers and are important parameters that would affect the rheological behaviour and sensory attributes of foods fortified with recrystallized starches.

3.5. Particle size distribution

Fig. 5 shows the particle size distributions of the recrystallized glucan powders and the median particle size and surface area to volume ratio are presented in Table 3. Depending on the acid reaction time of the starch prior to debranching and the recrystallization method, mono- and bi-modal particle size distributions are apparent. Acid depolymerization time of up to 96 h increased particle size of the DS powders whereas prolonged depolymerization (216 h) decreased the average particle size (Table 3). Furthermore, the proportion of particles comprising the distribution peak around 2–5 μm significantly decreased with increase in acid reaction time of the granular starch, indicating that larger aggregates were formed due to the absence of entangling long chain polymer networks (Gidley & Bulpin, 1989; Potocki-Veronese et al., 2005). Recrystallization of DS enhanced aggregation in the order: ANN > TC > HMT, which is evident from the median particle sizes (Table 3) and distribution curves (Fig. 5). This implies that HMT, in comparison to ANN and TC, induces stronger intermolecular forces such as those involving hydrogen bonds leading to higher stabilization of the glucans at the supra-molecular level. However, a more definitive parameter for assessing particle aggregation is the surface area to volume ratio. Assuming uniformity in density and shape, an inverse correlation exists between particle size and surface area (Lowell, Shields, Thomas, & Thommes, 2004). This rule is well obeyed in the glucan powders derived from the non- and 24 h acid-modified starches. The deviation from this rule in glucan powders derived from 96 and 216 h acid-modified starch could probably be due to differences in porosity, geometrical form and surface imperfections of the particles although surface area measurements based on particle size distribution are, at best, estimates of the real surface (Lowell et al., 2004).

3.6. *In vitro* digestibility

The *in vitro* digestibility kinetic curves of recrystallized glucan powders are shown in Fig. 6 and the kinetic parameters and starch digestion fractions are presented in Table 4. The R^2 values of the fitted curves were above 0.97 indicating that the observed data were adequately represented. Digestion proceeded rapidly up to about 60 min, and thereafter, slowly either due to a declining activity of the enzymes, an inhibition of the enzymes by hydrolysis products, declining substrate accessibility or a combination of these factors. Acid modification of the starch prior to debranching and the method of recrystallization of the linear glucans significantly ($p < 0.001$) influenced the digestion kinetics as can be seen from

Table 3
Median particle size (μm) and surface area to volume ratio (m^2/cm^3) of debranched and recrystallized starch powders.

	Median particle size				Surface volume ratio			
	0 h[A]	24 h	96 h	216 h	0 h	24 h	96 h	216 h
DS	$10.12^{a,B}_{v}$	14.45^{b}_{v}	45.62^{d}_{v}	25.29^{c}_{v}	1.00^{c}_{y}	0.75^{b}_{y}	0.39^{a}_{x}	0.43^{a}_{v}
ANN-DS	15.27^{a}_{x}	18.65^{b}_{x}	45.73^{d}_{v}	28.72^{c}_{x}	0.77^{d}_{x}	0.65^{c}_{x}	0.34^{a}_{y}	0.51^{b}_{x}
TC-DS	18.70^{a}_{y}	18.24^{a}_{x}	53.21^{c}_{x}	35.07^{b}_{y}	0.73^{c}_{x}	0.69^{c}_{x}	0.42^{a}_{x}	0.51^{b}_{x}
HMT-DS	37.09^{a}_{z}	38.82^{a}_{z}	67.62^{c}_{z}	40.73^{a}_{y}	0.64^{d}_{v}	0.55^{c}_{v}	0.27^{a}_{v}	0.40^{b}_{v}

DS: debranched, ANN-DS: debranched and annealed, TC-DS: debranched and temperature-cycled, HMT-DS: debranched and heat-moisture treated starch.
[A] Acid treatment time of the granular starch; 0 h represents the untreated starch.
[B] Means in the same column sharing similar subscript letters or in the same row sharing similar superscript letters do not differ significantly ($p < 0.05$).

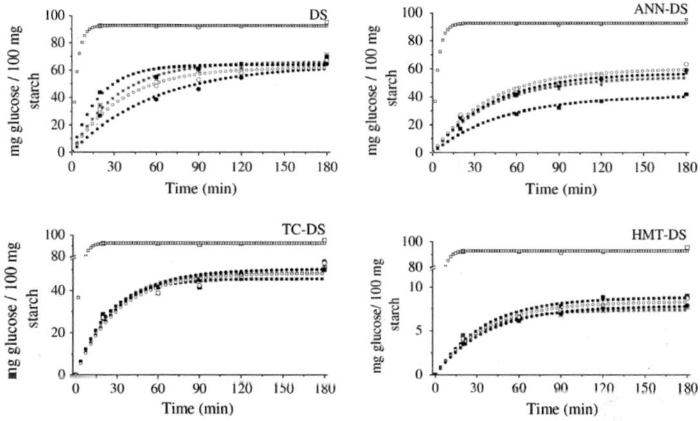

Fig. 6. In vitro digestibility kinetic curves of native (■), and 24 h (∗), 96 h (●), and 216 h (○) acid-modified debranched (DS), debranched and annealed (ANN-DS), debranched and temperature-cycled (TC-DS) and debranched and heat-moisture treated (HMT-DS) cassava starch powders. Gelatinized freeze-dried cassava starch powder (□) is the control.

Table 4
In-vitro digestibility kinetic parameters and nutritional fractions (g/100 g) of debranched and recrystallized starch powders.

	DS				ANN-DS				TC-DS				HMT-DS				Control[D]
	0 h[A]	24 h	96 h	216 h	0 h[A]	24 h	96 h	216 h	0 h[A]	24 h	96 h	216 h	0 h[A]	24 h	96 h	216 h	
k (1/min)[B,E]	5.2^{d}, [C]	3.2^{c}	1.7^{a}	2.4^{b}	2.0^{a}	2.3^{ab}	2.4^{b}	2.5^{b}	4.5^{d}	3.8^{a}	3.3^{a}	3.3^{a}	2.7^{a}	2.9^{a}	2.7^{a}	2.7^{a}	27.8
$C_{\infty\ (t=180)}$[E]	63.9^{a}	66.0^{b}	65.5^{ab}	65.8^{b}	40.8^{a}	55.4^{b}	56.9^{b}	60.0^{c}	45.6^{a}	48.4^{b}	50.1^{b}	48.2^{b}	8.7^{ab}	7.5^{a}	7.8^{a}	8.4^{a}	92.6
d_r (%)	62.8^{g}	59.9^{f}	48.5^{d}	54.9^{e}	32.8^{b}	46.3^{cd}	48.1^{d}	51.3^{e}	43.6^{c}	45.5^{c}	45.8^{c}	44.2^{c}	7.8^{a}	6.8^{a}	6.9^{a}	7.4^{a}	100.0
RDS	42.1^{f}	34.1^{e}	27.5^{c}	30.7^{d}	22.6^{b}	26.8^{c}	29.2^{d}	30.4^{d}	28.6^{cd}	28.9^{d}	26.7^{c}	29.6^{d}	4.3^{a}	3.9^{a}	3.5^{a}	4.0^{a}	92.3
SDS	22.4^{e}	28.5^{f}	30.4^{fg}	29.2^{f}	20.2^{d}	22.8^{e}	22.3^{e}	23.7^{e}	17.1^{bc}	17.4^{c}	20.6^{d}	14.8^{b}	4.2^{a}	3.3^{a}	4.0^{a}	3.7^{a}	0.0
RS	35.5^{a}	37.5^{b}	43.9^{d}	39.5^{c}	56.9^{i}	50.4^{f}	48.4^{e}	47.1^{e}	54.4^{h}	53.8^{g}	52.5^{g}	55.6^{hi}	91.4^{j}	92.7^{j}	92.5^{j}	92.2^{j}	7.3

DS: debranched, ANN-DS: debranched and annealed, TC-DS: debranched and temperature-cycled, HMT-DS: debranched and heat-moisture treated starch.
k: hydrolysis rate constant, C_∞: equilibrium reducing sugar content at $t = 180$ min, t: hydrolysis time (min), d_r: relative digestibility.
[A] non-acid treated.
[B] $\times 10^{-2}$.
[C] Means in the same row sharing similar superscript letters do not differ significantly ($p < 0.05$).
[D] Gelatinized freeze-dried cassava starch powder.
[E] Means are compared within and not across recrystallization method categories.

the C_∞ and k values on Table 4. Using the velocity descriptor k, the amylolytic susceptibilities of similarly recrystallized glucans can be compared. Acid degradation of the granular starch prior to debranching significantly ($p < 0.001$) decreased the enzyme susceptibility of DS. During debranching, the linear glucans are capable of re-associating into structures that resistant enzyme hydrolysis (Pohu, Planchot, Putaux, Colonna, & Buleon, 2004a). The long amylose chains, however, experience or induce mobility constraints and aggregation of the optimal glucans into stable crystalline entities is impeded. Consequently, partially cross-linked matrices that exhibit low resistance to amylolysis are formed. Thus the decreasing susceptibility of DS to enzyme hydrolysis with increasing acid treatment time is due to the enhanced ability of optimal glucans to retrograde as the long polymers become

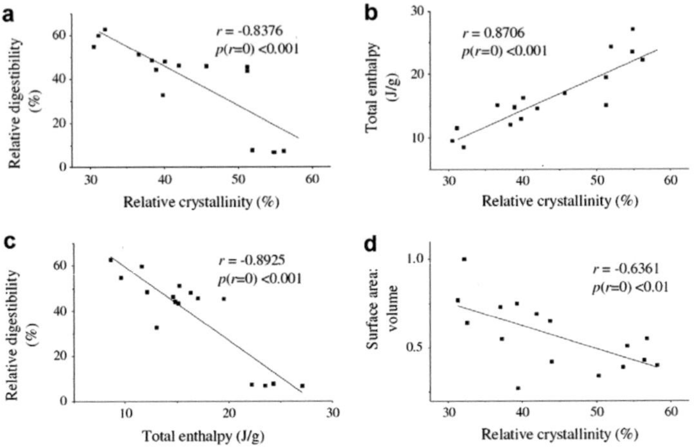

Fig. 7. Linear correlations between relative crystallinity, melting enthalpy, digestibility and surface area volume ratio of debranched and recrystallized cassava starch powders.

progressively depolymerized. Contrastingly, acid depolymerization had no significant ($p > 0.05$) influence on enzyme susceptibilities of ANN-DS, TC-DS or HMT-DS implying that ANN, TC and HMT minimized the structural variations that were responsible for the disparities in the initial hydrolysis kinetics among the DS powders. The effect of ANN, TC and HMT on recrystallization of linear glucans were discussed elsewhere (Mutungi, Rost, Onyango, Jaros, & Rohm, 2009). During hydrothermal incubation, realignment of the linear glucans takes place and molecular interactions leading to completion of double helix formation and growth of stable crystalline aggregates occur. A more inclusive parameter, however, for comparing the treatment effects across the recrystallization methods is relative digestibility, d_r. As evident in Table 4, the average d_r for the different recrystallization methods followed the order: DS > ANN-DS, TC-DS > HMT-DS and a significant ($p < 0.05$) interaction effect between acid depolymerization and recrystallization method was observed. The RDS, SDS, and RS fractions of the crystalline glucan powders are also presented in Table 4. The highest RDS contents were determined in the DS but significantly decreased with acid modification up to 96 h acid reaction time, which is consistent with the observed decreasing susceptibility to enzyme hydrolysis. The content of SDS, on the contrary, increased significantly ($p < 0.05$) with acid treatment as did the RS fraction in agreement with increasing relative crystallinity (Table 1).

RDS represents the amorphous component of the starch whereas SDS comprises the imperfect-crystallites consisting of amorphous material caged within crystalline structures (Guraya et al., 2001; Robin, Mérinat, Simon, & Lehmann, 2008; Shin et al., 2004). The results indicate that limited acid depolymerization of starch prior to debranching facilitates more ordered retrogradation as debranching progresses. On the other hand, no significant acid depolymerization effects could be assigned to the RDS and SDS contents in ANN-DS, TC-DS and HMT-DS although these were significantly lower than in DS, an observation consistent with enhanced polymer realignment and crystallization induced by these hydrothermal treatments (Lehmann et al., 2002; Mutungi et al., 2009). The RS contents decreased significantly ($p < 0.05$) with increasing acid depolymerization time in the ANN-DS powders but did not differ in the TC-DS or HMT-DS powders and the mean RS contents followed the order: HMT-DS > TC-DS > ANN-DS > DS suggesting that the recrystallization conditions induced discrete polymer interactions and the chain length distributions of the starches probably played a minor role when extreme conditions are applied.

Numerous intrinsic and extrinsic factors such the extent of polymer interactions, distribution and perfection of crystalline regions, branching structure, molecular weight, particle size, solubility, bound moisture and morphology influence the digestibility of recrystallized starch (Al-Rabadi, Gilbert, & Gidley, 2009; Pohu, Putaux, Planchot, Colonna, & Buleon, 2004b; Roder et al., 2009. Double helix formation is regarded as the basis for the resistance of recrystallized starches to α-amylase hydrolysis because the structures are too large to enter the active site and must be disentangled for catalytic conversion to occur (André, Buléon, Haser, & Tran, 1999). Following additional processing structural alterations take place and these influence further the rate of diffusion and adsorption of the enzyme on the solid substrate or the accommodation of the starch in the active sites of the enzymes (Colonna et al., 1992; Leloup, Colonna, & Ring, 1991).

Some linear correlations between structural parameters and relative digestibility of the glucan powders are presented in Fig. 7. The significant negative correlation between digestibility and relative crystallinity (Fig. 7a) supports the argument that crystalline structure hinders the accessibility of glucosidic bonds by amylolytic enzymes. However, high crystallinities also coincided with allomorphic transitions culminating in higher amounts of the A allomorph which, traditionally, is regarded as being more susceptible to enzyme hydrolysis (Planchot, Colonna, & Buleon, 1997; Williamson et al., 1992). Pohu et al. (2004b) have, however, reported high resistance of A-type recrystallized maltodextrins to amylolysis and attributed this property to its dense compact morphology arising from the epitaxial growth of elementary crystalline A-type platelets. The significant positive correlation between melting enthalpy and relative crystallinity (Fig. 7b) is an indicator of enhanced stabilization of double helices in the crystalline domains, and hence, the inverse correlation between enthalpy and digestibility (Fig. 7c). Furthermore, crystallinity was inversely correlated to the surface area volume ratio (Fig 7d). These relationships indicate that recrystallization induced formation of structures that hindered digestibility by different mechanisms. Formation of dense supra-molecular structures during recrystallization would exhibit steric incompatibility with the active site of α-amylase, and necessitate that high activation energies be overcome by the enzyme to adsorb to the substrate. The varying amylolysis susceptibilities of the recrystallized starch powders further indicate diverse bonding interactions which result in varied degrees of cohesion between the double helical strands and structural configurations that probably also correspond to diverse amylolysis hindrance mechanisms. On the other hand, depending on the starch and the digestion factors, polymer reorganisation may occur, presumably resulting in alteration of the initial hydrolysis kinetics (Htoon et al., 2009; Lopez-Rubio, Flanagan, Gilbert, & Gidley, 2008b). Moreover, the structure and behaviour of the non-crystalline fraction may also influence enzyme susceptibility, meaning that the correlations depicted in Fig. 7 are to be considered with caution.

4. Conclusion

Molecular size of the base starch and recrystallization conditions influenced the crystalline structure, thermal stability, and resistance of the recrystallized glucans to enzyme hydrolysis. Mild acid modification of cassava starch prior to debranching increased the kinetics of recrystallization and further processing, depending on the conditions, enhanced molecular interactions of the linear glucans. Increased crystallinity enhanced the thermal stability and enzyme resistance of the glucans. The degree of crystallinity is, however, to be taken as a measure of the density of packing of helical structures and may not necessarily reflect the detailed helical conformation and orientation of the non-crystalline fraction, which may exert significant confounding effects on particle size distribution, melting, solubility, water binding and digestibility of the recrystallized starch. The results provide practical approaches to tailor-make functional α-D-$(1 \rightarrow 4)$ glucans for use in a wide variety of food commodities. Further work is, nevertheless, needed to elucidate the short-range structure and the changes which may occur in the supra-molecular organization of the recrystallized starch during digestion as such structural alterations relate more closely to the physiological functionality of the starch.

Acknowledgements

Christopher Mutungi would like to thank Deutscher Akademischer Austauschdienst (DAAD) for the financial support. The authors thank Dr Albena Lederer and Susanne Boye of Leibnitz Institute of Polymer Research, Dresden for the analysis of molecular weight.

References

Adebowale, K. O., Henle, T., Shwarzenbolz, U., & Doert, T. (2009). Modification and properties of African yam bean (Sphenostylis stenocarpa Hochst Ex A. Rich.) harms starch I: Moisture treatments and annealing. *Food Hydrocolloids, 23*, 1947–1957.

Al-Rabadi, J. S. G., Gilbert, R. G., & Gidley, M. J. (2009). Effect of particle size on kinetics of starch digestion in milled barley and sorghum grains by porcine alpha-amylase. *Journal of Cereal Science, 50*, 198–204.

André, G., Buléon, A., Haser, R., & Tran, V. (1999). Amylose chain behavior in an interacting context. III. Complete occupancy of the AMY2 barley alpha-amylase cleft and comparison with biochemical data. *Biopolymers, 50*, 751–762.

Aparicio-Saguilán, A., Flores-Huicochea, E. J., Tovar García-Suárez, F., Gutiérrez-Meraz, F., & Bello-Pérez, L. A. (2005). Resistant starch-rich powders prepared by autoclaving of native and lintnerized banana starch: Partial characterization. *Starch/Stärke, 57*, 405–412.

Ares, G., Baixauli, R., Sanz, T., Varela, P., & Salvador, A. (2009). New functional fibre in milk puddings. Effect on sensory properties and consumers' acceptability. *LWT-Food Science and Technology, 42*, 710–716.

Arimi, J. M., Duggan, E., O'Riordan, E. D., O'Sullivan, M., & Lyng, J. G. (2008). Microwave expansion of imitation cheese containing resistant starch. *Journal of Food Engineering, 88*(2), 254–262.

Atichokudomchai, N., Chinachoti, P., & Varavinit, S. (2004). A study of ordered structure in acid-modified tapioca starch by ^{13}C CP/MAS solid-state NMR. *Carbohydrate Polymers, 58*, 383–389.

Baixauli, R., Salvador, A., Martinez-Cervera, S., & Friszman, S. M. (2008). Distinctive sensory features introduced by resistant starch in baked products. *LWT-Food Science and Technology, 41*, 1927–1933.

Belitz, H.-D., & Grosch, W. (1999). *Food chemistry*. Berlin: Springer.

Colonna, P., Leloup, V., & Buleon, A. (1992). Limiting factors of starch hydrolysis. *European Journal of Clinical Nutrition, 46*, S17–S32.

Dubois, M., Gilles, A. K., Hamilton, J. K., & Rebers, P. A. (1956). Colorimetric method for determination of sugars and related substances. *Analytical Chemistry, 28*, 350–356.

Eerlingen, R. C., Crombez, M., & Delcour, J. A. (1993). Enzyme-resistant starch. I. Quantitative and qualitative influence of incubation time and temperature of autoclaved starch on resistant starch formation. *Cereal Chemistry, 70*, 339–344.

Eerlingen, R. C., & Delcour, A. (1995). Formation, analysis, structure and properties of type III enzyme-resistant starch. *Journal of Cereal Science, 22*, 129–138.

Englyst, H. N., Kingman, S. M., & Cummings, J. H. (1992). Classification and measurement of nutritionally important starch fractions. *European Journal of Clinical Nutrition, 46*, 33–50.

Gelenscér, T., Gál, V., Hódsági, M., & Salgó, A. (2008). Evaluation of quality and digestibility characteristics of resistant starch-enriched pasta. *Food Bioprocess Technology, 1*, 171–179.

Gidley, M. J., & Bulpin, P. V. (1989). Aggregation of amylose in aqueous systems: The effect of chain length on behavior and aggregation kinetics. *Macromolecules, 22*, 341–346.

Gidley, J., Cooke, D., Darke, A. H., Hoffmann, R. A., Russel, A. L., & Greenwell, P. (1995). Molecular order and structure in enzyme-resistant retrograded starch. *Carbohydrate Polymers, 28*, 23–31.

Gõni, I., Garcia-Alonso, A., & Saura-Calixito, F. (1997). A starch hydrolysis procedure to estimate glycemic index. *Nutrition Research, 17*, 427–437.

González-soto, R. A., Mora-Escobedo, R., Hernández-Sánchez, H., Sánchez-Rivera, M., & Bello Pérez, L. A. (2007). The influence of time and storage temperature on resistant starch formation from autoclaved debranched banana starch. *Food Research International, 40*, 304–310.

Gunaratne, A., & Hoover, R. (2002). Effect of heat-moisture treatment on the structure and physicochemical properties of tuber and root starches. *Carbohydrate Polymers, 49*, 425–437.

Guraya, H. S., James, C., & Champagne, E. T. (2001). Effect of cooling and freezing on the digestibility of debranched rice starch and physical properties of the resulting material. *Starch/Stärke, 53*, 64–74.

Han, J.-A., BeMiller, J. N., Hamaker, B., & Lim, S.-T. (2003). Structural changes of debranched corn starch by aqueous heating and stirring. *Cereal Chemistry, 80*, 323–328.

Hoover, R. (2000). Acid treated starches. *Food Reviews International, 16*, 369–392.

Htoon, A., Shrestha, A. K., Flanagan, B. M., Lopez-Rubio, A., Bird, A. R., Gilbert, E. P., et al. (2009). Effects of processing high amylose maize starches under controlled conditions on structural organization and amylase digestibility. *Carbohydrate polymers, 75*, 236–245.

Jayakody, L., & Hoover, R. (2002). The effect of lintnerization on cereal starch granules. *Food Research International, 35*, 665–680.

Jayakody, L., & Hoover, R. (2008). Effect of annealing on the molecular structure and physicochemical properties of starches from different botanical sources – A review. *Carbohydrate Polymers, 74*, 691–703.

Le Bail, P., Bizot, H., & Buléon, A. (1993). B to A type phase transition in short chain amylose chains. *Carbohydrate Polymers, 21*, 99–104.

Lehmann, U., Jacobasch, G., & Schmiedl, D. (2002). Characterization of resistant starch type III from banana (*Musa acuminata*). *Journal of Agricultural and Food Chemistry, 50*, 5236–5240.

Lehmann, U., Rössler, C., Schmiedl, D., & Jacobasch, G. (2003). Production and physicochemical characterization of resistant starch type III from pea starch. *Nahrung/Food, 47*, 60–63.

Leloup, V. M., Colonna, P., & Ring, S. G. (1991). Alpha-amylase adsorption on starch crystallites. *Biotechnology and Bioengineering, 38*, 127–134.

Lopez-Rubio, A., Flanagan, M. B., Gilbert, P. E., & Gidley, J. M. (2008b). A novel approach for calculating starch crystallinity and its correlation with double helix content: A combined XRD and NMR study. *Biopolymers, 89*, 671–768.

Lopez-Rubio, A., Flanagan, B. M., Shrestha, A. K., Gidley, M. J., & Gilbert, E. P. (2008a). Molecular rearrangement of starch during in vitro digestion: Toward a better understanding of enzyme resistant starch formation in processed starches. *Biomacromolecules, 9*, 1951–1958.

Lowell, S., Shields, J. E., Thomas, M. A., & Thommes, M. (2004). *Characterization of porous solids and powders: Surface area, pore size and density*. Dordrecht: Kluwer Academic Publishers.

Moorthy, S. N. (2002). Physico-chemical and functional properties of tropical tuber starches: A review. *Starch/Stärke, 54*, 559–592.

Mutungi, C., Onyango, C., Jaros, D., Henle, T., & Rohm, H. (2009). Determination of optimum conditions for enzymatic debranching of cassava starch and synthesis of resistant starch type III using central composite rotatable design. *Starch/Stärke, 61*, 367–376.

Mutungi, C., Rost, F., Onyango, C., Jaros, D., & Rohm, H. (2009). Crystallinity, thermal and morphological characteristics of resistant starch type III produced by hydrothermal treatment of debranched cassava starch. *Starch/Stärke, 61*, 634–645.

Nelson, N. (1944). A photometric adaptation of the Somogyi method for the determination of glucose. *Journal of Biological Chemistry, 153*, 375–380.

Noronha, N., O'Riordan, E. D., & O'Sullivan, M. (2007). Replacement of fat with functional fibre in imitation cheese. *International Dairy Journal, 17*, 1073–1082.

Nugent, A. P. (2005). Health properties of resistant starch. *Nutrition Bulletin, 30*, 27–54.

Onyango, C., & Mutungi, C. (2008). Synthesis and in vitro digestion of resistant starch type III from enzymatically hydrolysed cassava starch. *International Journal of Food Science and Technology, 43*, 1860–1865.

Planchot, V., Colonna, P., & Buleon, A. (1997). Enzymatic hydrolysis of α-glucan crystallites. *Carbohydrate Research, 298*, 319–326.

Pohu, A., Planchot, V., Putaux, J. L., Colonna, P., & Buleon, A. (2004a). Split crystallization during debranching of maltodextrins at high concentration by isoamylase. *Biomacromolecules, 5*, 1792–1798.

Pohu, A., Putaux, J.-L., Planchot, V., Colonna, P., & Buleon, A. (2004b). Origin of the limited α-amylolysis of debranched maltodextrins crystallized in the A form: A TEM study on model substrates. *Biomacromolecules, 5*, 119–125.

Potocki-Veronese, G., Putaux, J.-L., Dupeyre, D., Albenne, C., Remaud-Siméon, M., Monsan, P., et al. (2005). Amylose synthesized in vitro by amylosucrase: Morphology, structure and properties. *Biomacromolecules, 6*, 1000–1011.

Robin, F., Mérinat, S., Simon, A., & Lehmann, U. (2008). Influence of chain length on α-1, 4-D-glucan recrystallization and slowly digestible starch formation. *Starch/Stärke, 60*, 551–558.

Roder, N., Gerard, C., Verel, A., Bogracheva, T. Y., Hedley, C. L., Ellis, P. R., et al. (2009). Factors affecting the action of α-amylase on wheat starch: Effects of water availability. An enzymic structural study. *Food Chemistry, 113*, 471–478.

Roger, P., & Colonna, P. (1992). The influence of chain length on the hydrodynamic behaviour of amylose. *Carbohydrate Research, 227*, 73–83.

Sajilata, M. G., & Singhal, R. S. (2005). Specialty starches for snack foods. *Carbohydrate Polymers, 59*, 131–135.

Sanz, T., Salvador, A., Baixauli, R., & Fiszman, S. M. (2009). Evaluation of four types of resistant starch in muffins. II. Effects in texture, colour and consumer response. *European Food Research and Technology, 22*, 197–204.

Schmiedl, D., Bäuerlein, M., Bengs, H., & Jacobasch, G. (2000). Production of heat stable butyrogenic resistant starch. *Carbohydrate Polymers, 43*, 183–193.

Shin, S. I., Choi, H. J., Chung, K. M., Hamaker, B., Park, K. H., & Moon, T. W. (2004). Slowly digestible starch from debranched waxy sorghum starch: Preparation and properties. *Cereal Chemistry, 81*, 404–408.

Singh, V., & Ali, S. Z. (2000). Acid degradation of starch. The effect of acid and starch type. *Carbohydrate Polymers, 41*, 191–195.

Tattiyakul, J., Pradipasena, P., & Asavasaksakul, S. (2007). Taro Colocasia esculenta (L.) schott amylopectin structure and its effect on starch functional properties. *Starch/Stärke, 59*, 342–347.

Tester, R. F., Karkalas, J., & Qi, X. (2004). Starch-composition, fine structure and architecture. *Journal of Cereal Science, 39*, 151–165.

Williamson, G., Belshaw, N. J., Self, D. J., Noel, T. R., Ring, S. G., Cairns, P., et al. (1992). Hydrolysis of A- and B-type crystalline polymorphs of starch by α-amylase and glucoamylase 1. *Carbohydrate Polymers, 18*, 179–187.

Yoo, S.-H., & Jane, Y.-L. (2002). Molecular weights and gyration radii of amylopectins determined by high performance size exclusion chromatography equipped with multi-angle laser-light scattering and refractive index detectors. *Carbohydrate Polymers, 49*, 307–314.

Yue, P., & Waring, S. (1998). Resistant starch in food applications. *Cereal Foods World, 43*, 690–695.

Appendix 5

Mutungi, C., Onyango, C., Doert, T., Paasch, S., Thiele, S., Machill, S., Jaros, D., & Rohm, H. (2011a)

Long- and short-range structural changes of recrystallised cassava starch subjected to *in vitro* digestion.

Food Hydrocolloids, **25**, 477-485.

Food Hydrocolloids 25 (2011) 477–485

Contents lists available at ScienceDirect

Food Hydrocolloids

journal homepage: www.elsevier.com/locate/foodhyd

Long- and short-range structural changes of recrystallised cassava starch subjected to *in vitro* digestion

C. Mutungi [a,*], C. Onyango [b], T. Doert [c], S. Paasch [c], S. Thiele [c], S. Machill [c], D. Jaros [a], H. Rohm [a]

[a] *Institute of Food Technology and Bioprocess Engineering, Technische Universitaet Dresden, D-01062 Dresden, Germany*
[b] *Max Rubner Institute, Schützenberg 12, D-32756 Detmold, Germany*
[c] *Department of Chemistry and Food chemistry, Technische Universitaet Dresden, D-01062 Dresden, Germany*

ARTICLE INFO

Article history:
Received 30 April 2010
Accepted 22 July 2010

Keywords:
Cassava starch
Resistant starch type III
In vitro digestion
Crystallinity
Molecular order
Structural conformations

ABSTRACT

The conformational and structural order of resistant starch type III made from cassava starch was studied. The gelatinized starch was debranched using pullulanase and then recrystallised by annealing, temperature-cycling or heat-moisture treatment. Subsequently, the recrystallised products were subjected to *in vitro* digestion using porcine pancreatic α-amylase and amyloglucosidase. The undigested and digested products were analyzed for polymer chain distribution, crystallinity, molecular order, structural conformations and thermal stability using high performance anion exchange chromatography, wide angle X-ray diffraction, Fourier transform infrared spectroscopy, ^{13}C CP/MAS nuclear magnetic resonance and differential scanning calorimetry, respectively. Average degree of polymerisation increased from 20 to 22 glucose units upon digestion. Both the undigested and digested starches comprised mixtures of A, B and V crystalline types. Percentage of crystallinities by X-ray diffraction were 40.9%, 50.7% and 56.2% in annealed, temperature-cycled and heat-moisture treated starches, respectively. These values increased to 47.9%, 54.4% and 58.2%, respectively, in the digested products. The ordered fractions in the undigested annealed, temperature-cycled and heat-moisture treated starches were 69.3%, 71.4% and 79.2%, respectively, as determined by ^{13}C CP/MAS nuclear magnetic resonance. However, the disordered phase was indistinct in the digested products although the contents of non-crystalline conformations were significantly ($p < 0.01$) higher. The melting enthalpies of the digested residues increased by factors of 2.50 in annealed, 2.53 in temperature-cycled and 2.06 in heat-moisture treated starches, suggesting molecular rearrangement in a manner related to the enzyme susceptibility of the initial materials.

1. Introduction

Processed starch is able to recrystallise into structures that resist digestion in the small intestine (Eerlingen & Delcour, 1995). This enzyme resistant starch (RS) arrives in the lower gut where part of it is fermented by gut microflora into short chain fatty acids and other substances that exhibit metabolic and colonic health benefits (Nugent, 2005). Four types of RS are distinguished: physically inaccessible starch (RS I), raw starch (RS II), retrograded or recrystallised starch (RS III), and chemically modified starch (RS IV) (Eerlingen & Delcour, 1995). RS III is structurally more stable and may thus be used as a functional fibre ingredient in foods (Sharma, Yadav, & Ritika, 2008). Furthermore, recrystallised starch has been found to possess superior bifidogenic and butyrogenic properties (Lehmann, Jacobasch, & Schmiedl, 2002; Martin, Dumon, Lecannu, & Champ, 2000; Rideout, Liu, Wood, & Fan, 2008; Robertson et al., 2001; Zampa et al., 2004).

Functionality, nutritional value, and the health implications of recrystallised starch depend on the conformation and physicochemical properties (Bird, Brown, & Topping, 2000; Topping, Fukushima, & Bird, 2003). Structural stability also appears to influence interactions between the starch and intestinal microbiota (Bird et al., 2000; Dongowski, Jacobasch, & Schmiedl, 2005; Jacobasch, Dongowski, Schmiedl, & Muller-Schmehl, 2006). Based on the starch type, the conformation and stability of recrystallised starch is contrived during processing and storage. Using small angle X-ray diffraction, Shamai, Shimoni, and Bianco-Peled (2004) and Zabar, Shimoni, and Bianco-Peled (2008) found that the main morphology comprises a B-polymorph with lamellas arranged in long-range periodicity when high amylose corn starch is processed and recrystallised at low temperature. On the other hand, a mixture of A- and V-polymorphs without defined periodicity was reported in the starch processed and recrystallised at high temperature. Differences in the prebiotic properties of the preparations (Lesmes, Beards, Gibson,

* Corresponding author. Tel.: +49 351 463 36943; fax: +49 351 463 37126.
E-mail address: chrismutungi@yahoo.co.uk (C. Mutungi).

0268-005X/$ – see front matter © 2010 Elsevier Ltd. All rights reserved.
doi:10.1016/j.foodhyd.2010.07.023

Tuohy, & Shimoni, 2008) indicate that morphological variations could be used to delineate the physiological capabilities of recrystallised starch.

In recent years, recrystallised short linear glucans were shown to exhibit a relatively high resistance to digestion. The structural basis for the high defiance to amylolysis was reported to be the formation of a dense compact morphology (Pohu, Putaux, Planchot, Colonna, & Buleon, 2004). A number of studies have disclosed the influence of structure on enzyme kinetic parameters, and the physicochemical and techno-functional properties of some of the RS III preparations (Lehmann, Rossler, Schmiedl, & Jacobasch, 2003; Lehmann et al., 2002; Mutungi et al., 2010). The physiological importance, was further proven by a high fermentability into short-chain fatty acids, particularly butyrate (Brouns et al., 2007; Dongowski et al., 2005; Lehmann et al., 2002; Schmiedl, Bauerlein, Bengs, & Jacobasch, 2000; Schwiertz, Lehmann, Jacobasch, & Blaut, 2002). The physiological effects of recrystallised starch, however, relate more closely to the structure of the starch subsequent to digestion rather than before digestion. This knowledge is critical in the production of recrystallised starch products with a more predictable physiological functionality. The aim of this study was thus to investigate the long and short range structural changes that occur following the *in vitro* digestion of differently recrystallised linear α-D-(1 → 4) glucans derived from cassava starch.

2. Materials and methods

2.1. Production of recrystallised starch

Cassava starch was debranched using pullulanase (Sigma P2986, Sigma-Aldrich, Laborchemikalien GmbH, Steinheim, Germany) as described elsewhere (Mutungi et al., 2010). The debranched starch (DS) was recrystallised by annealing (ANN-DS), temperature-cycling (TC-DS) or heat-moisture treatment (HMT-DS). ANN-DS was prepared by autoclaving 20% w/w DS slurries at 121 °C for 15 min followed by incubation at 60 °C for 24 h in a water bath. The TC-DS samples (20% w/w) were prepared by three autoclaving (121 °C for 2 h) and incubation cycles (60 °C for 1 h in the first two cycles and 60 °C for 16 h in the third cycle) in a water bath. The HMT-DS was prepared by moistening freeze-dried DS to 20% w/w using deionized water and autoclaving at 120 °C for 2 h. All samples were freeze-dried and finely ground using a motor and pestle at completion of treatment.

2.2. In vitro digestion of recrystallised starch

The method described by Shamai, Bianco-Peled, and Shimoni (2003) was used with modifications. Recrystallised freeze-dried starch (1 g) were weighed into 50 mL centrifuge tubes and mixed with 25 mL sodium acetate–hydrochloric acid buffer, pH 5.2 containing 0.02% sodium azide. Amylase solution was prepared by suspending 1 g (24 U/mg) porcine pancreatic α-amylase (Sigma A3176) in 10 mL buffer and incubating the mixture at 37 °C for 10 min before recovering the supernatant by centrifugation at 1500 × g. The supernatant was mixed with amyloglucosidase (300 U/mL, Sigma A7095) in a ratio of 8:1 (v/v) just before use, and 4 mL enzyme mixture was transferred to the starch suspensions. The starch–enzyme mixtures were incubated horizontally in a shaking water bath (170 strokes/min) maintained at 37 °C for 16 h, centrifuged at 5000 × *g* for 10 min and the supernatant decanted. To deproteinise the sample, the pellet was washed twice with 0.1 mol/L phosphate buffer, pH 7.4, and re-suspended in 25 mL of the buffer containing 1 mL protease solution prepared by dissolving 16 mg of fungal protease (Sigma P2143) in 100 mL of the buffer. The mixture was incubated in a shaking water bath at 42 °C for 4 h. The undigested starch was recovered by centrifugation and freeze dried. Resistant starch content was determined gravimetrically as weight of freeze-dried pellet divided by weight of sample multiplied by 100.

2.3. Polymer chain distribution

The polymer chain length distribution of the undigested and digested products was determined using high performance anion exchange chromatography equipped with a pulsed amperometric detector as described elsewhere (Mutungi et al., 2010) with Kestopentaose as internal standard. Degree of polymerisation (DPn) was the number of glucose units per polymer chain. Areas under individual peaks were integrated using Metrodata IC Net 2.3 software (Metrohm AG, Herisau, Switzerland). The average degree of polymerisation (DPn_{av}) was calculated using the formula:

$$DPn_{av} = \frac{\sum_{i=6}^{48} DPn_i \cdot A_{DPn_i}}{\sum_{i=6}^{48} A_{DPn_i}}$$

where *A* is peak area expressed as a percentage.

2.4. Wide angle X-ray diffraction (XRD)

A Siemens D5000 (Bruker-AXS, Karlsruhe, Germany) X-ray diffractometer was used. Samples were scanned with Cu Kα radiation ($\lambda = 0.154$ nm) in the angular range $2\theta = 2.5–70°$ with step intervals of 0.01° and a scanning duration of 6.5 s on each step for enhanced peak resolution. The X-ray generator operating conditions were 40 kV and 25 mA. Amorphous starch, prepared by autoclaving a 10% (w/w) cassava starch–water suspension followed by freeze drying after freezing at −82 °C for 20 min, was scanned for estimation of background scattering. Diffractograms were processed using PeakFit® software version 4.12 (Seasolve Software Inc., CA; USA) for the determination of degree of crystallinity and the proportions of A, B and V crystalline polymorphs.

2.5. Fourier transform infrared spectroscopy (FTIR)

FTIR spectra were obtained on a Nicolet 210 FTIR spectrometer (Nicolet Instrument GmbH, Offenbach, Germany). About 2–5 mg sample were mixed with 400 mg potassium bromide, homogenized and compressed. For each sample, 100 scans were recorded at room temperature (22 ± 1 °C) in the wave number interval of 400–4000 cm^{-1} with a spectral resolution of 2 cm^{-1} and co-added. A background spectrum was recorded using potassium bromide, and subtracted from the sample spectrum. The spectral region 1200–800 cm^{-1}, which is sensitive to the conformational and crystalline state of starch (van Soest, Tournois, de Wit, & Vliegenthart, 1995), was truncated and baseline corrected by subtracting a two point linear function touching 1200 and 880 cm^{-1} and the resulting spectra normalised to unit area. Gaussian function peaks were fitted at approximately 1155, 1125, 1105, 1080, 1047, 1022, 995, 965, 925, 860, and 840 cm^{-1}. Band intensities were the integrated areas under the peaks. Spectra processing was done using Peakfit® version 4.12.

2.6. ^{13}C CP/MAS nuclear magnetic resonance (NMR)

Solid-state ^{13}C cross-polarization and magic angle spinning (CP/MAS) NMR experiments were performed on a Bruker Avance 300 spectrometer (Bruker BioSpin, Karlsruhe, Germany) operating at 75.48 MHz for the ^{13}C nucleus using a commercial double resonance 4 mm MAS NMR probe. Ramped $^{1}H-^{13}C$ cross-polarization was used. Sample spectra were accumulated with a spinning rate of 14 kHz at the magic angle (54.7°). Cross-polarization contact time of 4 ms with a recycle delay of 2 s was applied and spectra acquired with a line broadening of 10 Hz. At least 2000 scans were accumulated for each spectrum at room temperature. Spinal ^{1}H-decoupling was applied during signal acquisition to remove proton spin–spin interactions and improve spectra resolution. Spectra were referenced relative to Tetramethylsilane. For quantitative analysis, the total spectra were decomposed into amorphous and ordered phases by subtracting a scaled amorphous starch spectrum until zero intensity was attained at 84 ppm for the difference spectrum (Tan, Flanagan, Halley, Whittaker, & Gidley, 2007). The amorphous starch spectrum was recorded from freeze-dried gelatinized starch. Both the total and ordered sub-spectra were deconvoluted and the intensities of deconvolution peaks of C1 resonance were used to calculate the proportion of the ordered phase, and the relative contributions by double helix, single helix and other conformations. PeakFit® version 4.12 was used for spectra processing.

2.7. *Differential scanning calorimetry (DSC)*

A DSC Q100 (TA Instruments Inc, Eschborn, Germany) calibrated using indium was used. Duplicate samples (3.0 ± 0.1 mg) were weighed into hermetic aluminium pans, and 10 μL of deionized water was added. The pans were hermitically sealed and allowed to equilibrate overnight at ambient temperature. DSC runs were performed from 10 to 170 °C at heating rate of 10 °C/min with an empty pan as reference and the data analyzed using the manufacturer's software, Universal Analysis 2000 V4.4.

2.8. *Statistical analysis*

Polymer chain distribution analysis, differential scanning calorimetry and resistant starch determinations were done in triplicates whereas XRD, FTIR, and ^{13}C CP/MAS NMR data were the outputs of spectral processing procedures involving three different manual inputs. Data were subjected to one-way ANOVA and means were separated using Fisher's least significant difference test on SigmaPlot® version 11.0 (Systat Software Inc., London, UK).

3. Results and discussion

3.1. *Resistant starch and polymer chain distribution*

Resistant starch contents of ANN-DS, TC-DS and HMT-DS were 62.46, 49.81 and 88.22 g/100 g, respectively. A discussion of the treatment methods and their effects on RS formation is provided elsewhere (Mutungi, Rost, Onyango, Jaros, & Rohm, 2009). A proposed mechanism for the resistance of recrystallised processed starch to amylolysis is the acquisition of a double-helical order because of the hindrance to enzyme binding that is brought about by these structures (Colonna, Leloup, & Buleon, 1992). Stabilization of the double helices into crystalline entities further lowers the vulnerability of the starch by decreasing the effective surface and thus the diffusion and adsorption of the enzyme onto the solid substrate. Entrapment of amorphous regions within imperfect crystals and non-crystalline double helical structures may also induce resistance to enzymatic digestion (Cairns, Sun, Morris, & Ring, 1995; Gidley et al., 1995).

Fig. 1 shows the polymer chain distributions of the recrystallised linear glucans before and after digestion. The average degree of polymerisations of ANN-DS, TC-DS and HMT-DS, were 20.46, 20.27, and 20.61 glucose units, respectively. Recrystallisation method did not affect ($p > 0.05$) the average polymer sizes, and the *in vitro* digestion residues had higher DPn of 22.96, 22.04 and 22.27, respectively. Furthermore, the amount of longer chains (DPn 18–44) increased in the digested starches (Fig. 1), suggesting that linear polymers capable of about 3–8 helical turns formed the more enzyme resistant structures. Eerlingen, Deceuninck, and Delcour (1993) observed that the complete alignment of longer polymers was impaired due to entropic reasons, and too short chains lacked the minimum length for formation of stable double helices. During digestion the long chains are also hydrolyzed into optimal-length chains which then acquire the capacity to form more stable enzyme resistant structures (Lopez-Rubio, Flanagan, Shrestha, Gidley, & Gilbert, 2008). The profile of the change in abundance of polymer chains was generally uniform for the digested recrystallised products and the maximum increases occurred at DPn ≈ 36 as shown in the lower row of Fig. 1. The magnitude differences, which follow the order ANN-DS > TC-DS > HMT-DS, reflect differences in enzyme hydrolysis kinetics of the starches that were earlier reported (Mutungi et al., 2010). Moreover, the local maxima at DPn 14 and DPn 20 in the polymer chain distribution of digested TC-DS indicate a bimodal polymer spread emanating from structurally dissimilar domains in the product. Other workers have reported DPn of 20–30 (Gidley, et al., 1995), 19–26 (Eerlingen et al., 1993), and 13–17 (Lopez-Rubio, Flanagan, Shrestha, et al., 2008) in resistant starch isolated from recrystallised starches of different origins. The observation that polymer chains of DPn ≈ 36 exhibit highest resistance to digestion agrees well with the mean DPn of 35 reported by Faisant et al. (1993) in the main resistant starch fraction isolated from the ileal terminus of the human small intestine following ingestion of retrograded starch.

3.2. *Crystallinity*

The X-ray diffractograms of the recrystallised starches and the digested residues are presented in Fig. 2a. The evolution patterns and intensities of reflections at $2\theta \approx 10°, 11°, 15°, 17°, 18°, 20°, 23°, 26°, 30°$, and 33° indicate significant qualitative and quantitative differences in crystallinities. The starches are mixtures of mainly A and B polymorphs, and small amounts of the V type polymorph. The A polymorph is normally distinguished by strong reflections at $2\theta = 17.14°$, 18.14°, 15.11° and 26.27°, with some additional reflections at $2\theta = 9.98°$, 11.19°, 22.93°, 23.68°, 30.30° and 33.08°, whereas the B polymorph is identified by typical reflections at $2\theta = 5.51°$, 14.60° and 16.85° with auxiliary peaks at $2\theta = 10.01°$, 11.02°, 13.85°, 22.30°, 23.71°, 26.16°, 30.61° and 33.84°. The V polymorph is recognized by a typical peak at: $2\theta = 19.80°$ and, sometimes, additional peaks occur at $2\theta = 7.40°$ and 12.90° (Lopez-Rubio, Flanagan, Gilbert, & Gidley, 2008; Godet, Bizot, & Buleon, 1995). To quantify the relative amounts of the different polymorphs, normalised sample and amorphous starch diffractograms were fitted with a linear baseline between $2\theta = 5–37.5°$, and the amorphous starch scattering scaled using an appropriate factor so as to fit under the minima points marked y and z, on Fig. 2b of the sample diffractogram, without generating negative intensities on the difference sub-diffractogram. An example of the crystalline fraction sub-diffractogram modelled with 12 Gaussian function peaks at $2\theta \approx 10°, 11°, 14°, 15°, 17°, 18°, 20°, 23°, 24°, 26°, 30°$ and 33° is shown in Fig. 2c. The R^2 of the modelled sub-diffractograms was at least 0.99 and the actual positions and intensities of peaks were used to calculate the relative amounts of A-, B-, and V-crystalline polymorphs by minimising the sum of square error between the observed and the respective standard peak positions (A polymorph: $2\theta = 9.98°$, 11.19°, 15.11°, 17.14°, 18.14°, 22.93°, 23.68°, 26.27°, 30.30°, 33.08°; B polymorph: $2\theta = 5.51°$, 10.01°, 11.02°,

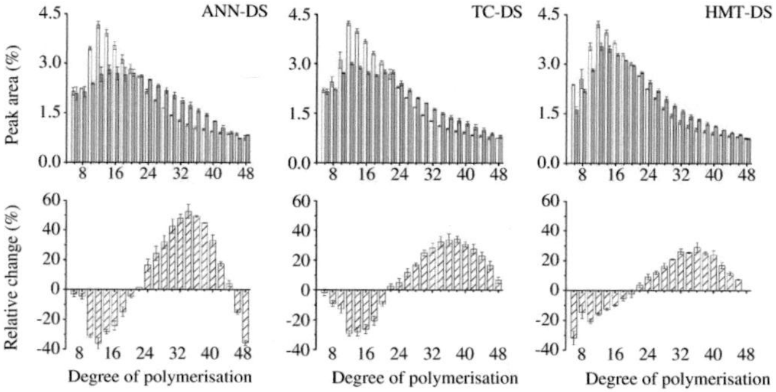

Fig. 1. Polymer chain distributions of recrystallised cassava starches before (□) and after (■) *in vitro* digestion (upper row). Starches were recrystallised by annealing (ANN-DS), temperature cycling (TC-DS) or heat-moisture treatment (HMT-DS). Lower row represents the relative increase or decrease in abundance of the polymer chains following *in vitro* digestion.

13.85°, 14.60°, 16.85°, 22.30°, 23.71°, 26.16°, 30.61°, 33.84°; V polymorph: $2\theta = 7.40°, 12.90°, 19.80°$). The degree of crystallinity of the starches expressed as percent fraction of the area of the crystalline sub-diffractogram relative to the area of the total diffractogram, and the amounts of A-, B- and V-polymorphs expressed as the percent fraction of the sum of peak areas corresponding to each of the polymorph patterns relative to the area of the total diffractogram are given in Table 1. Temperature cycling (TC-DS) and heat-moisture treatment (HMT-DS) favoured A-type crystalline packing, whereas a higher amount of B-type packing was evident in the ANN-DS due to the milder recrystallisation conditions. Similarly, V-type crystalline material was significantly higher in the TC-DS and HMT-DS suggesting that the higher temperature conditions enabled stabilization of a significantly higher amount of single helical chains into entities capable of diffracting X-rays.

The XRD traces of the digested starches show loss of reflection intensities at $2\theta \approx 15°, 18°$ and $26°$. This suggests a decline in A-type crystalline packing, and coincides with a decrease in the size of crystallites, or the formation of crystallographic dislocations within crystalline lattices as implied by the general broadening of reflections. These changes appear to be dependent on the crystalline rigidity of the undigested starch. By evaluating the XRD diffractograms on the basis of peak intensities referenced to a smooth curve connecting the individual peak minima as in the traditional two-phase method, it is apparent that the crystallinity degree of the digested starches would be lower. The two-phase approach assumes that samples comprise perfect crystals interspersed in a non-crystalline background, which, in processed and also in native starches, is not the case (Lopez-Rubio, Flanagan, Gilbert, et al., 2008). The alternative procedure of fitting the amorphous diffractogram (Fig. 2b) assumes a crystal-defect situation, where crystalline domains, instead, contain imperfections (Lopez-Rubio, Flanagan, Gilbert, et al., 2008). Using crystal-defects approach, the crystallinities of the digested materials were significantly higher than in the undigested starches (Table 1). This observation is explained by the fact that X-ray reflections correspond to the contributions of the long-range three-dimensional crystalline domains leaving out the contributions of irregularly packed crystallites, small chain aggregates and isolated single helices (Abd Karim, Norziah, & Seow, 2000; Lopez-Rubio, Flanagan, Gilbert, et al., 2008). Moreover, the intensity of crystalline peaks is subject to the stability of atoms in their ideal positions and deviations may lead to loss of peak intensities. The higher crystallinities in the digested materials coincided with higher amounts of B-type packing

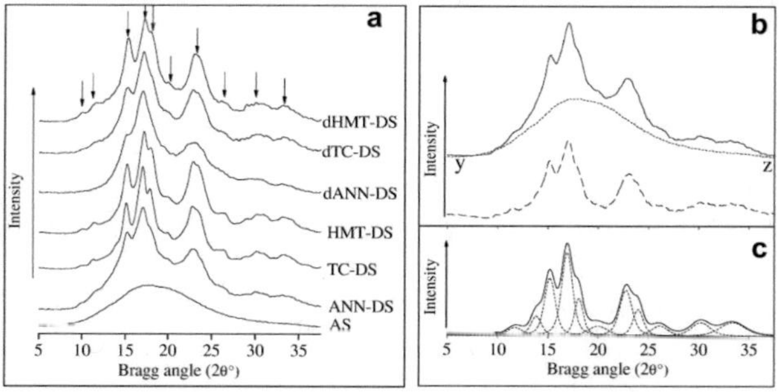

Fig. 2. Wide-angle X-ray diffractograms: (a) diffractograms of amorphous (AS), and the recrystallised cassava starches before and after digestion. Starches were recrystallised by annealing (ANN-DS), temperature cycling (TC-DS) or heat-moisture treatment (HMT-DS). Digested starches are additionally marked with the letter 'd'. Arrows show the evolution of reflection peaks; (b) sample diffractogram fitted with amorphous starch diffractogram. Ends marked 'y' and 'z' represent fit reference points; and (c) modelled crystalline fraction sub-diffractogram fitted with Gaussian function peaks.

Table 1
Crystallinity, molecular order and melting parameters of recrystallised cassava starch before and after digestion.

Sample	XRD^{X}				FTIR^{Y}		NMR^{X}					DSC	
	C_d	A_p	B_p	V_p	1047/1022	995/1022	O_{fr}	C_{dh}	NC_{dh}	Cd_{cf}	V_{cf}	$T_c - T_o$ (°C)	ΔH (J/g)
AS	nd	nd	nd	nd	0.40^{a}	0.46^{g}	nd	nd	nd	nd	nd	nd	nd
ANN-DS	40.90$^{a\ Z}$	23.33^{a}	14.33^{c}	3.24^{a}	1.03^{a}	1.61^{a}	69.34^{a}	44.78^{a}	9.39^{c}	6.50^{c}	8.67^{c}	40.58^{a}	14.98^{b}
TC-DS	50.70^{b}	40.09^{b}	7.01^{b}	3.81^{c}	1.35^{b}	3.03^{b}	71.43^{b}	55.33^{b}	4.31^{a}	3.25^{a}	8.54^{b}	71.34^{c}	13.08^{a}
HMT-DS	56.18^{c}	52.02^{c}	0.75^{a}	3.42^{b}	2.17^{c}	3.70^{c}	79.23^{c}	61.78^{c}	7.32^{b}	5.25^{b}	4.88^{a}	46.07^{b}	24.38^{c}
dANN-DS	47.88$_{a}$	12.58$_{a}$	33.44$_{c}$	1.87$_{a}$	4.51$_{b}$	9.09$_{b}$	nd	49.87$_{a}$	15.67$_{c}$	8.68$_{a}$	25.78$_{c}$	69.72$_{b}$	37.43$_{b}$
dTC-DS	54.41$_{b}$	38.70$_{c}$	13.13$_{b}$	2.70$_{b}$	4.37$_{a}$	8.33$_{a}$	nd	56.85$_{b}$	10.96$_{b}$	10.27$_{b}$	21.92$_{b}$	68.15$_{b}$	33.07$_{a}$
dHMT-DS	58.19$_{c}$	48.77$_{d}$	6.99$_{a}$	2.43$_{c}$	5.95$_{c}$	10.00$_{c}$	nd	64.87$_{c}$	8.79$_{a}$	8.54$_{a}$	17.80$_{a}$	70.31$_{b}$	50.28$_{c}$

AS: amorphous starch. Recrystallised starches were prepared by annealing (ANN-DS), temperature cycling (TC-DS) or heat-moisture treatment (HMT-DS). The digested starches are additionally marked with the letter 'd'.
C_d: degree of crystallinity; A_p: A-polymorph; B_p: B-polymorph, V_p: V-polymorph; O_{fr}: ordered fraction; C_{dh}: crystalline double helices; NC_{dh}: non-crystalline double helices; Cd_{cf}: constrained conformations; V_{cf}: V-type conformations; $T_c - T_o$: melting temperature range; ΔH: change in enthalpy.
nd: Not determined.
[X] Percent proportion.
[Y] Intensity ratios of bands representative for crystallinity (1047/1022) and molecular order (995/1022).
[Z] Values in the same column having same superscript (undigested) or subscript (digested) are not significantly different ($p < 0.05$).

and a concurrent decrease in the A- and V-type crystalline packing. The magnitudes by which B-morphologies increased or A- and V-morphologies decreased as a result of digestion was higher in the ANN-DS and lower for TC-DS and HMT-DS (Table 1). This observation is consistent with the development of less dense double-helical packing and suggests hydrolysis-related loss of structural rigidity. Furthermore, increase in crystallinity was significantly higher in ANN-DS as compared to TC-DS and HMT-DS suggesting that the amount of structural reorganization during digestion was related to the structural rigidity of the starting material.

3.3. *Molecular order*

Infrared spectroscopy is sensitive to the oscillatory frequencies of chemical bonds; vibrational modes generating absorption bands at approximately 1155, 1125, and 1105 cm^{-1} (CO, CC stretching with some COH contributions), 1080, 1047, 1022, 995, and 928 cm^{-1} (COH bending and CH_2 related modes), and 860 cm^{-1} (COC symmetrical stretching and CH deformation) were shown to correspond to the conformational and crystalline order of starch (van Soest, et al., 1995). FTIR was thus used to probe the molecular order of the undigested and digested recrystallised starches. The band peaks at ~995 and ~1047 cm^{-1} relate to molecular order and crystallinity of starch polymers whereas vibrational modes resulting in the band intensity at ~1022 cm^{-1} correspond to the disordered or amorphous phase (Capron, Robert, Colonna, Brogly, & Planchot, 2007; Lopez-Rubio, Flanagan, Shrestha, et al., 2008; Htoon et al., 2009). FTIR spectra of the starches are shown in Fig. 3. Among the undigested starches, the intensity of the band at ~1022 cm^{-1} decreased in the order ANN-DS > TC-DS > HMT-DS. This absorption band, however, was not evident in the digested starches thus the amorphous conformations, if present, were contained in insignificant amounts in these materials. The spectral band peaks at ~1047 and ~995 cm^{-1}, and the intervening valley at ~1035 cm^{-1} also show patterns of interest. The trough at ~1035 cm^{-1}, in particular, increased in the order ANN-DS < TC-DS < HMT-DS, and showed greater evolution in the digested starches. The gradual evolution of this particular spectral valley coincides with the narrowing of the absorption band at ~1047 cm^{-1}, and depicts a narrower distribution of bond energies due to reduction in the number of varied conformations (van Soest, de Wit, Tournois, & Vliegenthart, 1994). The band peak at ~1047 cm^{-1} is sensitive to crystallinity thus the narrowing of this band is due to ordering of the starch into crystalline double helices.

Intensities of the bands at ~1047 cm^{-1} and ~995 cm^{-1} increased in the order ANN-DS < TC-DS < HMT-DS among the undigested starches (Fig. 3). In the digested starches, however, the band at ~995 cm^{-1} also shifted to a higher wave number i.e. ~1001 cm^{-1}. A similar band shift has been reported to occur upon hydration of partially crystalline starch (Capron et al., 2007; van Soest et al., 1995) and is expected because of the structuring effect of water on starch polymers, which further causes the bands at ~995 cm^{-1} and ~1022 cm^{-1} to overlap. However, Sevenou, Hill, Farhat, and Mitchell (2002) also observed that acid hydrolysis of the amorphous fraction of granular starch caused the band at ~1022 cm^{-1} to weaken, and due to its dominance, the band at ~995 cm^{-1} shifts to a higher wave number. The absorption band located at ~995 cm^{-1} is ascribed to single helix crystalline structure environment (van Soest et al., 1995), thus predominance of the band at ~1000 cm^{-1} in the digested starch implies an increase in the amount of single helices attaining molecular order due to intramolecular hydrogen bonding.

The ratios of the integrated areas of absorption bands i.e. 1047/1022 cm^{-1} and 995/1022 cm^{-1} were used to compare the degree of crystalline and molecular order of the starches (Table 1). Use of ratios instead of the individual band intensities allows compensation for

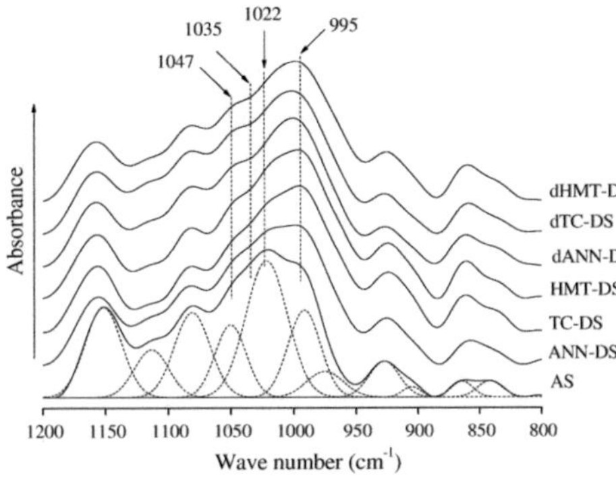

Fig. 3. Fourier transform infrared spectra of amorphous starch (AS) and the recrystallised cassava starches before and after digestion. Starches were recrystallised by annealing (ANN-DS), temperature cycling (TC-DS) or heat-moisture treatment (HMT-DS). Digested starches are additionally marked with the letter 'd'. Example of peaks fitted for DS is shown at the bottom.

vibration mode overlap effects. Among the undigested starches, crystallinity and molecular order were highest in HMT-DS and lowest in ANN-DS as indicated by the increasing 1047/1022 cm^{-1} and 995/1022 cm^{-1} band intensity ratios, respectively. Furthermore, molecular order is evidently higher in the digested materials although dTC-DS had significantly lower short-range order compared to dANN-DS and dHMTDS. This probably reflects variations in the degree of polymer rearrangement in response to digestion kinetics of the differently recrystallised starches. A similar change in the molecular order of starch following digestion was reported in high amylose starches extruded under different conditions (Lopez-Rubio, Flanagan, Shrestha, et al., 2008).

3.4. Structural conformations

Based on the uniqueness of the C1 resonance pattern, ^{13}C CP/MAS NMR has been used for structural organization investigations of starch (Gidley & Bociek, 1988; Morgan, Furneaux, & Larsen, 1995; Paris, Bizot, Emery, Buzare, & Buleon, 1999). Fig. 4a shows the ^{13}C CP/MAS NMR spectra of the recrystallised starches before and after digestion. The ratio between the combined intensities of C2, C3, C4 and C5 resonances and the intensity of C1 resonance was constant at 4.42 ± 0.20 for the different samples spectra. This ratio constancy provided confidence that data from the deconvoluted C1 signal could reliably be employed to quantitatively compare the structural order across the starches. In the C1 region (90–110 ppm) local electron densities relate more directly to glucosidic torsion angles thus peak patterns exhibit a more straight-forward correlation with the conformational order of the starch (Paris et al., 1999). As expected, the amorphous starch exhibits wide chemical shift dispersion with significant intensity peaks at 103, 100, 97 and 95 ppm (Fig. 4a). This peak register is attributable to a broad range of local chain conformations, characterized by glycosidic linkage torsion angles that define the relative orientation of adjacent glucose units within the polymer chains (Durran, Howlin, Webb, & Gidley, 1995; Gidley & Bociek, 1988).

The deconvolution data of C1 resonances are given in Table 2. ANN-DS resolved into a doublet at 101 and 100 ppm whereas TC-DS and HMT-DS resolved into a triplet at 101, 100 and 99 ppm. These peak multiplicity patterns were retained in the respective starches after digestion. A doublet or triplet peak pattern at these chemical shifts typifies B or A crystalline lattices, respectively, because in the B-polymorph, a maltose is the asymmetric repeat unit in each turn of the double helix strand as opposed to a maltotriose in the A-polymorph (Paris et al., 1999). The resonance peaks are, however, ambiguously resolved on the NMR traces (Fig. 4a). This could be because the starches are polymorphic mixtures and the helical packing symmetries overlap (Bogracheva, Wang, & Hedley, 2001), or the double helices are not sufficiently crystalline so as to yield narrow distributions of the glucosidic bond torsion angles associated with the respective chemical shifts (Horii, Yamamoto, Hirai, & Kitamaru, 1987). It can, however, be inferred that the observed peak multiplicities are inclined towards the dominant polymorph implying higher amounts of A-type crystallinity in HMT-DS and TC-DS as compared to ANN-DS.

To quantify the ordered and non-ordered conformations, the total spectra were decomposed based on the assumption that the spectra are linear combinations of the signatures of the two phases (Paris et al., 1999). Spectral decomposition of the undigested starch spectra with reference to the C4 resonance, whose intensity is solely due to amorphous contributions (Tan et al., 2007) is illustrated in Fig. 5. The C4 resonance decreased in the order ANN-DS > TC-DS > HMT-DS in line with declining contents of the amorphous phase. Furthermore the spectral fingerprints of ANN-DS, TC-DS and HMT-DS also differ in the sharpness of the C2, 3 and 5 envelope resonances at 71, 72 and 76 ppm probably reflecting further disparities in the structural rigidity of the samples (Gidley & Bociek, 1988; Tang & Hills, 2003). Among the digested starches, the C4 signal was indistinct. It is possible that hydration could weaken the resonance intensity of the C4 atom due to water-induced crystallisation or segmental mobility resulting in insufficient cross-polarization (Bogracheva et al., 2001; Cheetham & Tao, 1998; Morgan et al., 1995; Paris et al., 1999; Tang & Hills, 2003). As all samples were freeze dried and thereafter stored under identical conditions, the lack of a resonance signal at the C4 site in the digested samples is due to removal of the amorphous conformations during digestion. Closer evaluation of the digested starch spectra, however, shows that the C1 resonances are generally broader, although, the positions of resonance peaks were retained (Fig. 4a). This signifies slight deviation from the glucosidic bond torsion angles distributions of highly homogeneous rigid conformations. Due to enzyme mediated disentanglement of the starches, less rigid conformations were thus formed. Nevertheless, it was not possible to decompose the ^{13}CP/MAS NMR spectra of the digested starches.

Examples of the deconvoluted total and ordered spectra are presented in Fig. 4b and c. In the ordered fraction, the collective intensities at 99, 100, and 101 ppm relative to the total intensity of the C1 signal have been used to calculate the relative crystallinity of starch (Atichokudomchai, Varavinit, & Chinachoti, 2004; Lopez-Rubio,

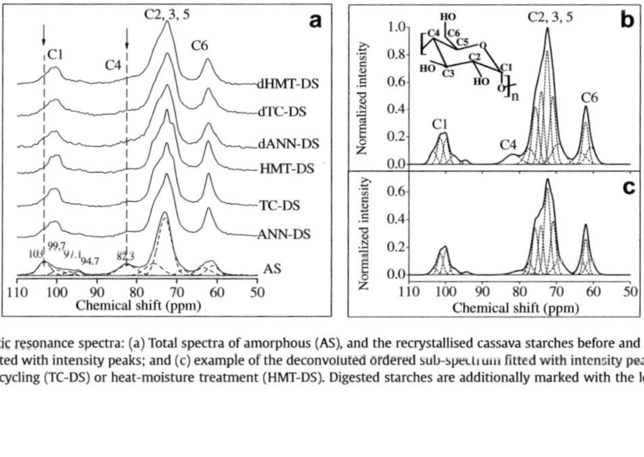

Fig. 4. ^{13}C CP/MAS nuclear magnetic resonance spectra: (a) Total spectra of amorphous (AS), and the recrystallised cassava starches before and after digestion; (b) example of the deconvoluted total NMR spectra fitted with intensity peaks; and (c) example of the deconvoluted ordered sub-spectrum fitted with intensity peaks. Starches were recrystallised by annealing (ANN-DS), temperature cycling (TC-DS) or heat-moisture treatment (HMT-DS). Digested starches are additionally marked with the letter 'd'.

Table 2
Intensity (i)[a] and width at half height (w)[b] of deconvoluted ^{13}C CP/MAS NMR C1 resonance peaks.

	Chemical shift (ppm)	ANN-DS		TC-DS		HMT-DS		dANN-DS		dTC-DS		dHMT-DS	
		i	*w*	*i*	*w*	*i*	*w*	*i*	*w*	*i*	*w*	*i*	*w*
Total spectra	95	0.14	230.21	0.09	228.70	0.12	209.08	0.12	187.95	0.15	239.27	0.11	211.34
	98	0.18	163.04	0.09	140.39	0.16	169.83	0.22	168.32	0.16	118.50	0.12	166.81
	99	nd		0.26	95.10	0.24	87.56	nd	nd	0.28	92.84	0.23	99.63
	100	0.39	129.07	0.29	95.10	0.28	92.84	0.36	127.56	0.27	92.09	0.32	101.90
	101	0.36	133.60	0.32	107.94	0.29	106.43	0.34	131.34	0.28	112.47	0.31	111.71
	103	0.30	193.23	0.23	154.73	0.21	187.95	0.36	186.44	0.32	174.36	0.24	169.08
Ordered sub-spectra	95	0.09	230.97	0.04	127.56	0.07	171.34	nd	nd	nd	nd	nd	nd
	98	0.13	167.57	0.06	128.32	0.1	134.35	nd	nd	nd	nd	nd	nd
	99	nd	nd	0.24	99.63	0.26	88.31	nd	nd	nd	nd	nd	nd
	100	0.31	122.28	0.23	109.45	0.28	89.07	nd	nd	nd	nd	nd	nd
	101	0.31	126.05	0.24	107.18	0.28	98.88	nd	nd	nd	nd	nd	nd
	103	0.12	153.22	0.11	144.17	0.07	117.75	nd	nd	nd	nd	nd	nd

Recrystallised starches were prepared by annealing (ANN-DS), temperature cycling (TC-DS) or heat-moisture treatment (HMT-DS). Digested starches are additionally marked with the letter 'd'.
[a] Integrated area.
[b] Hz; nd: Not detected.

Flanagan, Gilbert, et al., 2008; Paris et al., 1999) and thus these can be considered to be representative of the crystalline double-helix component. The peak at 103 ppm is unanimously assigned to mono-helical conformations (Cheetham & Tao, 1998; Gidley & Bociek, 1988; Horii et al., 1987; Morgan, Furneaux, & Stanley, 1992; Paris et al., 1999), whereas the spectral region between 93 and 98 ppm, was observed in the total spectra of native, amorphous and amylose inclusion complexes and attributed to non-ordered conformations (Gidley & Bociek, 1988; Morrison, Tester, Gidley, & Karkalas, 1993). In our samples, however, the contribution of the 93–98 ppm spectral region in the ordered phase was also significant and two resonance peaks were resolved at ~95 and ~98 ppm. Morgan et al. (1995), Paris, Bizot, Emery, Buzare, and Buleon (2001) and Tang & Hills (2003) assigned the peak at ~95 ppm to constrained conformations, whereas the peak at ~98 ppm was reported as an exotic case of twisted glucosidic linkages in α-cyclodextrin (Gidley & Bociek 1988) and later correlated to α-D-(1 → 6) linkages in granular and amorphous starches (Morgan et al., 1992; Paris et al., 2001). The presence of this fraction in the debranched starches implies that some linear polymers may exist as distorted species. As the widths of the deconvoluted peak at ~98 ppm particularly in the total spectra are intermediate between those of the crystalline double helices and amorphous or mono-helical conformations (Table 2), a glucosidic bond torsion angles distribution that is intermediate between the crystalline and the amorphous or mono-helical states is implied. Moreover, the intensities at ~98 ppm and also ~95 ppm are lower in TC-DS compared to HMT-DS and ANN-DS meaning that the conformations are a consequence of energy and therefore mobility constraints. Thus in the ordered spectra, the peak at ~98 ppm comprises conformations that closely relate to the mainstream morphologies but are incapable of eliciting signals at similar chemical shifts as the crystalline double helices because of the energetically weak intra-molecular interactions i.e. non-crystalline double helices.

Ordered fractions in the undigested material increased in the order ANN-DS < TC-DS < HMT-DS (Table 1) but with varying amounts of crystalline double helices, non-crystalline double helices, constrained conformations and mono-helical chains. The amounts of crystalline double helices correspond well with XRD degree of crystallinity. The NMR values are, however, approximately 10% higher supporting the view that some crystalline material is less repetitive or irregularly packed and therefore unable to diffract X-rays (Cooke & Gidley, 1992). Furthermore, contents of the V-type (mono-helical) glucosidic conformations by XRD and NMR are not comparable because of the low amounts contained in the materials. The contents of mono-helical conformations were significantly lower in HMT-DS than in ANN-DS and TC-DS whereas TC-DS contained significantly lower amounts of non-crystalline double helices and other constrained conformations. These trends suggest that high temperature conditions favour mobilization of the polymers into double helices. The more stringent conditions involving high temperature and a considerable amount of plasticization further enhance mobilization of the energetically constrained conformations but the effect is less on the ordering of mono-helical chains.

The digested starches contained significantly higher amounts of the various conformations. Although double helical order increased, significant amounts of non-crystalline double helices, constrained and V-type conformations were also determined.

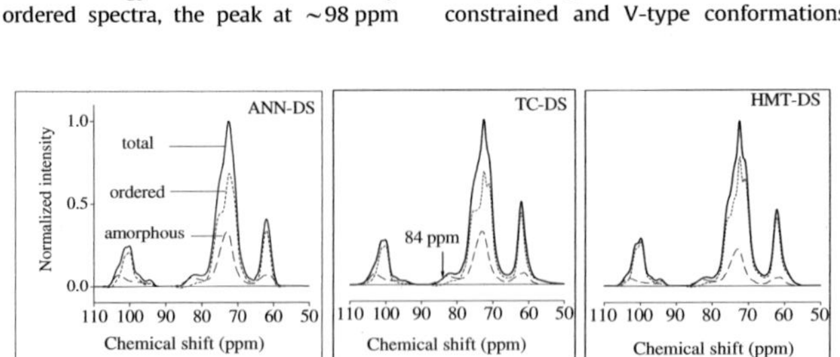

Fig. 5. ^{13}C CP/MAS NMR spectra decomposition of the recrystallised starches into ordered and disordered phase contributions. Starches were recrystallised by annealing (ANN-DS), temperature cycling (TC-DS) or heat-moisture treatment (HMT-DS).

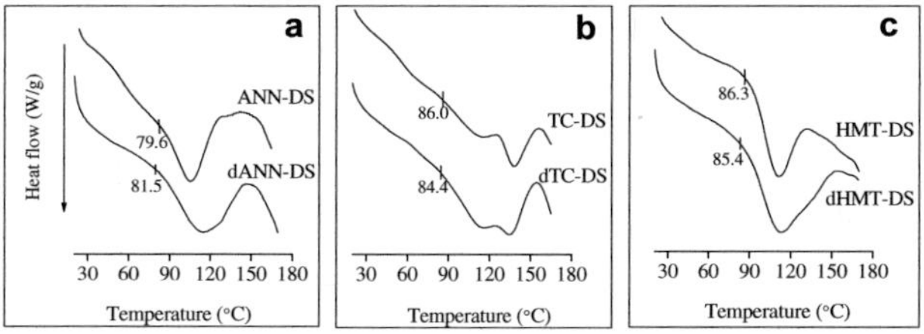

Fig. 6. Differential scanning calorimetry thermograms of recrystallised cassava starches before and after digestion. Recrystallised cassava starches were prepared by (a) annealing (ANN-DS), (b) temperature cycling (TC-DS) or (c) heat-moisture treatment (HMT-DS). Digested starches are additionally marked with the letter 'd'.

Furthermore, the latter were almost 10 times higher than the V-type crystallinity determined by XRD. Hence, some disruption of crystalline rigidity occurred during digestion in the order ANN-DS > TC-DS > HMT-DS. Nevertheless, as the FTIR data indicate, digestion coincided with higher degree of molecular order which suggests that a new starch colloidal structure was formed. In an investigation of the nanostructure of digested and non-digested extruded high amylose maize starch, Lopez-Rubio, Htoon, and Gilbert (2007) found an exotic repetitive arrangement in dry digested extruded high amylose starch. The arrangement was, however, not detected upon hydration as it was incorporated into the main periodic structure of the retrograded starch due to water-induced recrystallisation. Based on some unanimous observations on the changes in ^{13}CP/MAS NMR spectral features of the C1 signal following starch hydration (Bogracheva et al., 2001; Paris et al., 1999; Tang and Hills, 2003), it is reasonable to hypothesise that the non-crystalline conformations observed in our case subscribe to the exotic molecular organization described by Lopez-Rubio et al., (2007).

3.5. Thermal properties

DCS thermograms of the recrystallised starches before and after digestion are presented in Fig. 6, and the transition temperature range ($T_c - T_o$) and transition enthalpies (ΔH) have been summarised in Table 1. The effects of the different recrystallisation methods on the melting properties of the undigested recrystallised starches were discussed elsewhere in detail (Mutungi et al., 2009). The ($T_c - T_o$) can be interpreted as a measure of the homogeneity of crystalline and non crystalline conformations whereas ΔH reflects the amount of order i.e. the enthalpy of melting of crystallites and double helices (Abd Karim et al., 2000). Unlike ANN-DS and HMT-DS, the TC-DS exhibits two major endothermic peaks consistent with two distinct populations of crystalline domains induced by the high temperature cycling regime during preparation of the material. The kink at ~125 °C in the dTC-DS indicates that the individual domains were partly retained during digestion. Furthermore, the ($T_c - T_o$) was similar in both TC-DS and dTC-DS, whereas broadening of endothermic peaks characterized by shift of the transition conclusion temperatures (T_c) to higher levels occurred in dANN-DS and dHMT-DS (Fig. 6). Moreover, the transition peak temperature of dANN-DS shifted to a higher temperature implying that ANN-DS existed in a less cohesive state in comparison to TC-DS and HMT-DS and polymer rearrangement during digestion was more intense. The increases in ($T_c - T_o$) can be explained in terms of formation of a broader range of ordered arrays involving longer chains as implied by the relocation of T_c to higher temperature, and a more independent melting of individual crystallites and double helices subsequent to hydrolysis of the amorphous phase. The higher enthalpies further indicate attainment of higher molecular order as a consequence of removal of the disordered arrangements during digestion; the increases by factors of 2.50 in dANN-DS, 2.53 in dTC-DS and 2.06 in dHMT-DS is sufficient confirmation that molecular rearrangement occurred during digestion, but in a manner related to enzyme susceptibility of the initial material.

4. Conclusion

Because recrystallised starch is able to withstand digestion in the upper gut, a considerable amount reaches the lower gut where it may be fermented into short-chain fatty acids and other health promoting metabolites, besides supporting growth of beneficial gut microbiota and maintaining a lowered colonic pH. These physiological effects relate to the structure of the starch subsequent to digestion. In this study, conformational and structural order of recrystallised starch and that of the *in vitro* digestion residue were compared. The recrystallised starches comprised both ordered and amorphous phases whose relative abundances were dependent on the method of recrystallisation. Temperature cycling and heat-moisture treatment favoured A-type crystalline packing whereas the starch recrystallised by annealing exhibited higher proportion of the B-type morphology. Only small amounts of V-type crystalline packing were discernible in the starches. The ordered conformations could further be compartmentalised into crystalline double helices, non-crystalline double helices, constrained conformations and single helices; the relative contents were also dependent on the recrystallisation method. Although crystallinity plays a role in determining resistance of starch to digestion, amorphous conformations that are caged within imperfect crystals, non-crystalline double-helical structures, single helices, and constrained conformations may restrict enzyme hydrolysis in a manner related to their proximity to the crystalline regions. Following *in vitro* digestion, the crystallinity and molecular order of the starches were enhanced. Generally, A-type crystalline packing decreased whereas the B- and V-type morphologies increased and the disordered phase became indistinct. The digested starches were structurally less rigid, and the stoichiometry of the resultant conformations appears to be influenced by the structural rigidity of the initial material. The A-type morphology is apparently more resistant to structural alteration during digestion.

Acknowledgements

Christopher Mutungi would like to thank Deutscher Akademischer Austauschdienst (DAAD) for the financial support.

References

Abd Karim, A., Norziah, M. H., & Seow, C. C. (2000). Methods for the study of starch retrogradation. *Food Chemistry, 71*(1), 9–36.

Atichokudomchai, N., Varavinit, S., & Chinachoti, P. (2004). A study of ordered structure in acid-modified tapioca starch by C-13 CP/MAS solid-state NMR. *Carbohydrate Polymers, 58*(4), 383–389.

Bird, A. R., Brown, I. L., & Topping, D. L. (2000). Starches, resistant starches, the gut microflora and human health. *Current Issues in Intestinal Microbiology, 1*(1), 25–37.

Bogracheva, T. Y., Wang, Y. L., & Hedley, C. L. (2001). The effect of water content on the ordered/disordered structures in starches. *Biopolymers, 58*(3), 247–259.

Brouns, F., Arrigoni, E., Langkilde, A. M., Verkooijen, I., Fassler, C., Andersson, H., et al. (2007). Physiological and metabolic properties of a digestion-resistant maltodextrin, classified as type 3 retrograded resistant starch. *Journal of Agricultural and Food Chemistry, 55*(4), 1574–1581.

Cairns, P., Sun, L., Morris, V. J., & Ring, S. G. (1995). Physicochemical studies using amylose as an *in-vitro* model for resistant starch. *Journal of Cereal Science, 21*(1), 37–47.

Capron, I., Robert, P., Colonna, P., Brogly, M., & Planchot, V. (2007). Starch in rubbery and glassy states by FTIR spectroscopy. *Carbohydrate Polymers, 68*(2), 249–259.

Cheetham, N. W. H., & Tao, L. P. (1998). Solid state NMR studies on the structural and conformational properties of natural maize starches. *Carbohydrate Polymers, 36*(4), 285–292.

Colonna, P., Leloup, V., & Buleon, A. (1992). Limiting factors of starch hydrolysis. *European Journal of Clinical Nutrition, 46*, S17–S32.

Cooke, D., & Gidley, M. J. (1992). Loss of crystalline and molecular order during starch gelatinization – origin of the enthalpic transition. *Carbohydrate Research, 227*, 103–112.

Dongowski, G., Jacobasch, G., & Schmiedl, D. (2005). Structural stability and prebiotic properties of resistant starch type 3 increase bile acid turnover and lower secondary bile acid formation. *Journal of Agricultural and Food Chemistry, 53*(23), 9257–9267.

Durran, D. M., Howlin, B. J., Webb, G. A., & Gidley, M. J. (1995). Ab-initio nuclear shielding calculations of a model alpha-(1 → 4)-glucan. *Carbohydrate Research, 271*(2), C1–C5.

Eerlingen, R. C., & Delcour, J. A. (1995). Formation, analysis, structure and properties of type-III enzyme resistant starch. *Journal of Cereal Science, 22*(2), 129–138.

Eerlingen, R. C., Deceuninck, M., & Delcour, J. A. (1993). Enzyme-resistant starch .2. Influence of amylose chain-length on resistant starch formation. *Cereal Chemistry, 70*(3), 345–350.

Faisant, N., Champ, M., Colonna, P., Buleon, A., Molis, C., Langkilde, A. M., et al. (1993). Structural features of resistant starch at the end of the human small-intestine. *European Journal of Clinical Nutrition, 47*(4), 285–296.

Gidley, M. J., & Bociek, S. M. (1988). ^{13}C CP/MAS NMR Studies of amylose inclusion complexes, cyclodextrins, and the amorphous phase of starch granules: relationships between glycosidic linkage conformation and solid-state ^{13}C chemical shifts. *Journal of the American Chemical Society, 110*(12), 3820–3829.

Gidley, M. J., Cooke, D., Darke, A. H., Hoffmann, R. A., Russell, A. L., & Greenwell, P. (1995). Molecular order and structure in enzyme-resistant retrograded starch. *Carbohydrate Polymers, 28*(1), 23–31.

Godet, M. C., Bizot, H., & Buleon, A. (1995). Crystallization of amylose–fatty acid complexes prepared with different amylose chain lengths. *Carbohydrate Polymers, 27*(1), 47–52.

Horii, F., Yamamoto, H., Hirai, A., & Kitamaru, R. (1987). Structural study of amylose polymorphs by cross-polarization magic-angle spinning, ^{13}C NMR spectroscopy. *Carbohydrate Research, 160*, 29–40.

Htoon, A., Shrestha, A. K., Flanagan, B. M., Lopez-Rubio, A., Bird, A. R., Gilbert, E. P., et al. (2009). Effects of processing high amylose maize starches under controlled conditions on structural organization and amylase digestibility. *Carbohydrate Polymers, 75*(2), 236–245.

Jacobasch, G., Dongowski, G., Schmiedl, D., & Muller-Schmehl, K. (2006). Hydrothermal treatment of Novelose 330 results in high yield of resistant starch type 3 with beneficial prebiotic properties and decreased secondary bile acid formation in rats. *British Journal of Nutrition, 95*(6), 1063–1074.

Lehmann, U., Jacobasch, G., & Schmiedl, D. (2002). Characterization of resistant starch type III from banana (*Musa acuminata*). *Journal of Agricultural and Food Chemistry, 50*(18), 5236–5240.

Lehmann, U., Rossler, C., Schmiedl, D., & Jacobasch, G. (2003). Production and physicochemical characterization of resistant starch type III derived from pea starch. *Nahrung-Food, 47*(1), 60–63.

Lesmes, U., Beards, E. J., Gibson, G. R., Tuohy, K. M., & Shimoni, E. (2008). Effects of resistant starch type III polymorphs on human colon microbiota and short chain fatty acids in human gut models. *Journal of Agricultural and Food Chemistry, 56*(13), 5415–5421.

Lopez-Rubio, A., Flanagan, B. M., Gilbert, E. P., & Gidley, M. J. (2008>). A novel approach for calculating starch crystallinity and its correlation with double helix content: a combined XRD and NMR study. *Biopolymers, 89*(9), 761–768.

Lopez-Rubio, A., Flanagan, B. M., Shrestha, A. K., Gidley, M. J., & Gilbert, E. P. (2008). Molecular rearrangement of starch during *in vitro* digestion: toward a better understanding of enzyme resistant starch formation in processed starches. *Biomacromolecules, 9*(7), 1951–1958.

Lopez-Rubio, A., Htoon, A., & Gilbert, E. P. (2007). Influence of extrusion and digestion on the nanostructure of high-amylose maize starch. *Biomacromolecules, 8*(5), 1564–1572.

Martin, L. J. M., Dumon, H. J. W., Lecannu, G., & Champ, M. M. J. (2000). Potato and high-amylose maize starches are not equivalent producers of butyrate for the colonic mucosa. *British Journal of Nutrition, 84*(5), 689–696.

Morgan, K. R., Furneaux, R. H., & Larsen, N. G. (1995). Solid-state NMR-studies on the structure of starch granules. *Carbohydrate Research, 276*(2), 387–399.

Morgan, K. R., Furneaux, R. H., & Stanley, R. A. (1992). Observation by solid-state ^{13}C CP MAS NMR-spectroscopy of the transformations of wheat-starch associated with the making and staling of bread. *Carbohydrate Research, 235*, 15–22.

Morrison, W. R., Tester, R. F., Gidley, M. J., & Karkalas, J. (1993). Resistance to acid-hydrolysis of lipid-complexed amylose and lipid-free amylose in lintnerised waxy and non-waxy barley starches. *Carbohydrate Research, 245*(2), 289–302.

Mutungi, C., Onyango, C., Rost, F., Doert, T., Jaros, D., & Rohm, H. (2010). Structural and physicochemical properties and *in vitro* digestibility of recrystallized linear α-D-(1 → 4) glucans derived from mild-acid modified cassava starch. *Food Research International, 43*(4), 1144–1154.

Mutungi, C., Rost, F., Onyango, C., Jaros, D., & Rohm, H. (2009). Crystallinity, thermal and morphological characteristics of resistant starch type III produced by hydrothermal treatment of debranched cassava starch. *Starch-Starke, 61*(11), 634–645.

Nugent, A. P. (2005). Health properties of resistant starch. *Nutrition Bulletin, 30*, 27–54.

Paris, M., Bizot, H., Emery, J., Buzare, J. Y., & Buleon, A. (1999). Crystallinity and structuring role of water in native and recrystallized starches by ^{13}C CP-MAS NMR spectroscopy 1: spectral decomposition. *Carbohydrate Polymers, 39*(4), 327–339.

Paris, M., Bizot, H., Emery, J., Buzare, J. Y., & Buleon, A. (2001). NMR local range investigations in amorphous starchy substrates I. Structural heterogeneity probed by ^{13}C CP-MAS NMR. *International Journal of Biological Macromolecules, 29*(2), 127–136.

Pohu, A., Putaux, J. L., Planchot, V., Colonna, P., & Buleon, A. (2004). Origin of the limited alpha-amylolysis of debranched maltodextrins crystallized in the A form: a TEM study on model substrates. *Biomacromolecules, 5*(1), 119–125.

Rideout, T. C., Liu, Q., Wood, P., & Fan, M. Z. (2008). Nutrient utilization and intestinal fermentation are differentially affected by the consumption of resistant starch varieties and conventional fibres in pigs. *British Journal of Nutrition, 99*(5), 984–992.

Robertson, J. A., Ryden, P., Botham, R. L., Reading, S., Gibson, G., & Ring, S. G. (2001). Structural properties of diet-derived polysaccharides and their influence on butyrate production during fermentation. *Lebensmittel-Wissenschaft Und-Technologie-Food Science and Technology, 34*(8), 567–573.

Schmiedl, D., Bauerlein, M., Bengs, H., & Jacobasch, G. (2000). Production of heat-stable, butyrogenic resistant starch. *Carbohydrate Polymers, 43*(2), 183–193.

Schwiertz, A., Lehmann, U., Jacobasch, G., & Blaut, M. (2002). Influence of resistant starch on the SCFA production and cell counts of butyrate-producing *Eubacterium* spp. in the human intestine. *Journal of Applied Microbiology, 93*(1), 157–162.

Sevenou, O., Hill, S. E., Farhat, I. A., & Mitchell, J. R. (2002). Organization of the external region of the starch granule as determined by infrared spectroscopy. *International Journal of Biological Macromolecules, 31*(1–3), 79–85.

Sharma, A., Yadav, B. S., & Ritika. (2008). Resistant starch: physiological roles and food applications. *Food Reviews International, 24*(2), 193–234.

Shamai, K., Shimoni, E., & Bianco-Peled, H. (2004). Small angle X-ray scattering of resistant starch type III. *Biomacromolecules, 5*(1), 219–223.

Shamai, K., Bianco-Peled, H., & Shimoni, E. (2003). Polymorphism of resistant starch type III. *Carbohydrate Polymers, 54*(3), 363–369.

Tan, I., Flanagan, B. M., Halley, P. J., Whittaker, A. K., & Gidley, M. J. (2007). A method for estimating the nature and relative proportions of amorphous, single, and double-helical components in starch granules by ^{13}C CP/MAS NMR. *Biomacromolecules, 8*(3), 885–891.

Tang, H. R., & Hills, B. P. (2003). Use of ^{13}C MAS NMR to study domain structure and dynamics of polysaccharides in the native starch granules. *Biomacromolecules, 4*(5), 1269–1276.

Topping, D. L., Fukushima, M., & Bird, A. R. (2003). Resistant starch as a prebiotic and synbiotic: state of the art. *Proceedings of the Nutrition Society, 62*(1), 171–176.

van Soest, J. J. G., de Wit, D., Tournois, H., & Vliegenthart, J. F. G. (1994). Retrogradation of potato starch as studied by Fourier-transform infrared-spectroscopy. *Starch-Stärke, 46*(12), 453–457.

van Soest, J. J. G., Tournois, H., de Wit, D., & Vliegenthart, J. F. G. (1995). Short-range structure in (partially) crystalline potato starch determined with attenuated total reflectance Fourier-transform IR spectroscopy. *Carbohydrate Research, 279*, 201–214.

Zabar, S., Shimoni, E., & Bianco-Peled, H. (2008). Development of nanostructure in resistant starch type III during thermal treatments and cycling. *Macromolecular Bioscience, 8*(2), 163–170.

Zampa, A., Silvi, S., Fabiani, R., Morozzi, G., Orpianesi, C., & Cresci, A. (2004). Effects of different digestible carbohydrates on bile acid metabolism and SCFA production by human gut micro-flora grown in an *in vitro* semi-continuous culture. *Anaerobe, 10*(1), 19–26.

Appendix 6

Mutungi, C., Shuldt, S., Onyango, C., Schneider, Y., Jaros D., & Rohm, H. (2011c)

Dynamic moisture sorption characteristics of enzyme-resistant recrystallised cassava starch.

Biomacromolecules, **12**, 660-671.

BIOMACROMOLECULES

ARTICLE

pubs.acs.org/Biomac

Dynamic Moisture Sorption Characteristics of Enzyme-Resistant Recrystallized Cassava Starch

Christopher Mutungi,[*,†] Stefan Schuldt,[†] Calvin Onyango,[‡] Yvonne Schneider,[†] Doris Jaros,[†] and Harald Rohm[†]

[†]Institute of Food Technology and Bioprocess Engineering, Technische Universitaet Dresden, D-01069 Dresden, Germany

[‡]Food Technology Division, Kenya Industrial Research and Development Institute, P.O. Box 30650-00100, Nairobi, Kenya

ABSTRACT: The interaction of moisture with enzyme-resistant recrystallized starch, prepared by heat-moisture treatment of debranched acid-modified or debranched non-acid-modified cassava starch, was investigated in comparison with the native granules. Crystallinities of the powdered products were estimated by X-ray diffraction. Moisture sorption was determined using dynamic vapor sorption analyzer and data fitted to various models. Percent crystallinities of native starch (NS), non-acid-modified recrystallized starch (NAMRS), and acid-modified recrystallized starch (AMRS) were 39.7, 51.9, and 56.1%, respectively. In a_w below 0.8, sorption decreased in the order NS > NAMRS > AMRS in line with increasing sample crystallinities but did not follow this crystallinity dependence at higher a_w because of condensation and polymer dissolution effects. Adsorbed moisture became internally absorbed in NS but not in NAMRS and AMRS, which might explain the high resistance of the recrystallized starches to digestion because enzyme and starch cannot approach each other over fairly sufficient surface at the molecular level.

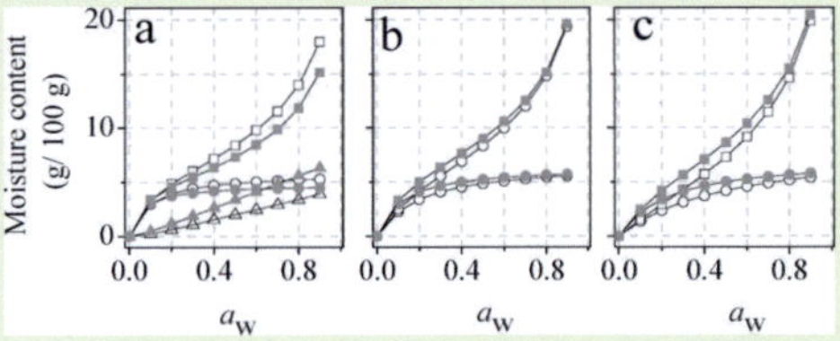

1. INTRODUCTION

Recrystallized starch, also known as resistant starch type III, is formed when native starch (NS) is gelatinized and allowed to retrograde.[1] The starch resists digestion in the small intestine, and reaches the lower gut, where part of it is fermented by gut microbiota into short-chain fatty acids and other substances that have metabolic and colonic health benefits.[2] For this reason, recrystallized starch may be used as a functional fiber ingredient in foods. In the majority of products such as baked foods,[3] imitation cheeses,[4] puddings,[5] and pasta,[6] where resistant starch inclusion has been proposed, product performance is related to the interaction of the starch polymers with moisture. Consequently, a majority of methods used in the production of this functional fiber involve a chemical, enzymatic, or thermal modification step not only to increase yield but also to modify the technofunctional suitability for use in a wider array of foods.[1] In combination with recrystallization conditions, the modified starch can rearrange into diverse spatial forms. The forces responsible for the rearrangements exert influence on characteristics such as crystallinity, phase and state transitions, gelling character, enzyme susceptibility, and other properties that are related directly to the state of water in the starch. From a technological point of view, inclusion of recrystallized starch in food matrices may also affect surface properties, and in solid products, for example, new networks of interlocking pores could form, thereby influencing the retention and migration of water in the product. Knowledge of interaction of recrystallized starch with moisture is therefore relevant not only to stability, packaging, and storage but also to the physical, textural, and sensory attributes of the enriched products and may additionally be vital in solving operational problems that might involve moisture transfer such as ingredients mixing.

Porous materials can retain water in "bound" and "free" states in the form of an adsorbed monomolecular layer or multilayers that produce pendular, funicular, or capillary bonding states between particles.[7] At low moisture contents, independent particles are wetted, and liquid films are formed on their surfaces. At the points where neighboring wetted particles contact each other, discrete liquid bridges may be formed. Surface tension and negative capillary pressure in such bridges provide the cohesive forces that result in the pendular state, and as the liquid content increases, several bridges coalesce, giving rise to the funicular state. Eventually, as more liquid is adsorbed, interstitial voids are filled, and the capillary state is attained. Moreover, the adsorbed moisture is subject to binding and diffusional forces and can be absorbed into the interior of particles when diffusional forces exceed the binding ones.[8]

The relationship between equilibrium moisture content of a material and the moisture level of its surrounding is described by a moisture sorption isotherm. This isotherm can be obtained in two ways, depending on whether the material is gaining (adsorption) or

Received: November 5, 2010
Revised: December 19, 2010
Published: January 24, 2011

losing moisture (desorption) at a specific temperature. Because thermodynamic equilibrium may not be achieved when the material is successively subjected to the two equilibration paths, a hysteresis characterizing the difference is usually observed, whose magnitude is related to a material's molecular components. Hysteresis reflects the extent to which these components undergo structural and conformational rearrangements, which alter the number of energetically favorable sites and migration of water within the material during the sorption processes.[9] To explain sorption behaviors of matrices, numerous kinetic, empirical, and semiempirical models were proposed by different workers and recently reviewed by several authors.[10,11] None of these models is universally expedient partially because the mechanisms by which water interacts with material surfaces vary enormously depending on the affinity for water as well as the pressure of water vapor within a material relative to the surrounding. The Brunnauer–Emmet–Teller (BET) equation is particularly widely used to simulate sorption phenomena.[11] Many researchers, however, modified this model to improve its relevance in the high water activity region. Some typical modifications include the Guggenheim–Anderson–de Boer (GAB) model[12] and the Timmermann's third sorption stage (TSS) simulation.[13]

The aim of this study was to investigate the interaction of moisture with enzyme-resistant recrystallized starch produced by heat-moisture treatment of debranched acid-modified and debranched non-acid-modified cassava starches by dynamic vapor sorption and to compare its interaction behavior with the interaction properties of the native granules.

2. MATERIALS AND METHODS

2.1. Modification and Recrystallization of Cassava Starch.

Native cassava starch was obtained from Kenya Industrial Research and Development Institute, Nairobi, Kenya. Acid-modified starch was prepared by suspending 40 g of the NS in 60 mL of 0.14 mol/L hydrochloric acid, followed by incubation at 40 °C for 9 days, after which the suspension was neutralized to pH 7 using 1 mol/L sodium hydroxide and centrifuged (1500*g*). The pellet was washed three times in excess deionized water before being dried in an oven at 40 °C. The molecular weight of the native and acid-modified starches was determined by asymmetric flow field flow fractionation, as detailed elsewhere.[14] The acid-modified and non-acid-modified starches were subsequently debranched using pullulanase (Sigma P2986, Sigma-Aldrich, Laborchemikalien GmbH, Steinheim, Germany), and the polymer chain length distributions of the debranched materials were determined using high-performance anion exchange chromatography equipped with a pulsed amperometric detector.[14] Debranched starches were further subjected to a heat-moisture treatment step,[14] then freeze-dried, ground using mortar and pestle, and passed through a 56 μm sieve. Samples for moisture sorption analysis were stored over phosphorus pentoxide in moisture tight Perspex chambers. Crystallinities of the products were determined using a Siemens D5000 (Bruker-AXS, Karlsruhe, Germany) X-ray diffractometer. Samples were scanned with Cu Kα radiation (λ = 0.154 nm) in the angular range 2θ = 2.5–37.5°. A step interval of 0.01° and a scanning duration of 6.5 s on each step were applied. The X-ray generator operating conditions were 40 kV and 25 mA. The degree of crystallinity was calculated as described elsewhere.[14] Resistant starch contents of the materials were determined gravimetrically as percent amount of starch remaining after incubation with a mixture of porcine pancreatic α-amylase (Sigma A3176) and amyloglucosidase (Sigma A7095) at 37 °C for 16 h.[14]

2.2. Dynamic Vapor Sorption Analysis.

A Q5000 SA dynamic vapor sorption analyzer (TA Instruments, Eschborn, Germany) equipped with an autosampler and microbalance capable of weighing with sensitivity of <0.1 μg was used. The desired relative humidity (RH) in the equilibration chamber is achieved through automated mixing of dry and moist nitrogen gas streams, and the humidified gas is allowed to flow continuously past the sample. An operating temperature of 25 °C was set, and a stepwise RH increment was programmed. Once a sample (~3 mg) was loaded, a 0% RH was maintained until the relative change in sample mass remained below 0.001% for 5 min, and the measurement was started. A single measurement protocol involved three cycles (Figure 1): a first stepwise adsorption from 0% RH to 90% RH, followed by a stepwise desorption to 0% RH at a step interval of 10%. Finally, a second stepwise adsorption with additional equilibration steps at RH of 94 and 98% was conducted. At each RH level, equilibration was stopped when the relative change in sample mass remained below 0.01% for 5 min, and the next RH step was automatically applied. Measurements were done in duplicate.

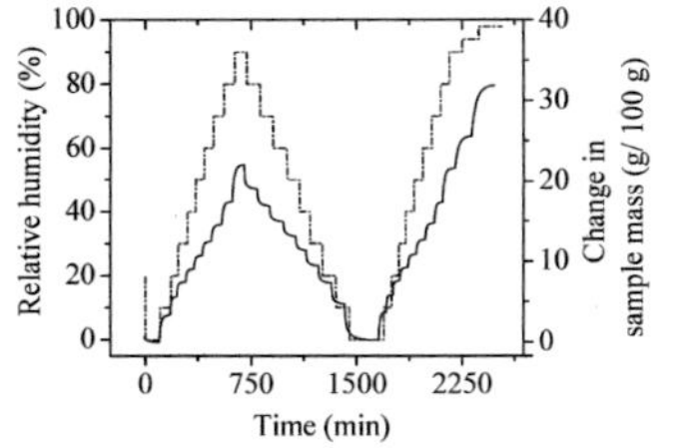

Figure 1. Stepwise increment or decrement in relative humidity (dashed line) and the associated change in sample mass (solid line) of a randomly selected starch sample during equilibration in dynamic vapor sorption analyzer.

2.3. Comparison of Sorption Isotherms and Model Fitting.

2.3.1. Sorption Hysteresis. The degree of hysteresis was calculated as the difference between desorption equilibrium moisture content (g/100 g) and adsorption equilibrium moisture content (g/100 g) expressed as a percent of the adsorption equilibrium moisture content.

2.3.2. Brunauer–Emmet–Teller Model. Sorption isotherms were fitted to the BET equation expressed as

$$M_{eq} = \frac{M_0 \times C_b \times a_w}{(1 - a_w)(1 + (C_b - 1)a_w)} \quad (1)$$

where M_{eq} is the equilibrium moisture load per dry matter at a given RH expressed as water activity, a_w, M_0 is the monolayer moisture load, and C_b is the BET monolayer energy constant. To calculate BET parameters, we rearranged eq 1 to a linear polynomial in terms of a_w[15] as follows

$$\frac{a_w}{(1 - a_w)M_{eq}} = \alpha + \beta \cdot a_w \quad (2)$$

where $\alpha = 1/(M_0 \cdot C_b)$ and $\beta = 1/(C_b - 1)/(M_0 \cdot C_b)$.

Least-squares regression of eq 2 was performed over the range of experimental points on OriginPro 8 (OriginLab Corporation, Northampton, MA). From the intercept (α) and gradient (β), M_0 and C_b were computed as $1/(\alpha + \beta)$ and $(\alpha + \beta)/\alpha$, respectively.

2.3.3. Guggenheim–Anderson–de Boer Model. The GAB model, an expanded form of the BET equation, estimates M_{eq} as

$$M_{eq} = \frac{M_0 \times C_g \times k_g \times a_w}{(1 - k_g a_w)(1 + (C_g - 1)k_g a_w)} \quad (3)$$

where C_g and k_g are characteristic constants corresponding to GAB monolayer and multilayer sorption energies, respectively. When k_g equals unity, eq 3 reverts to the BET model (eq 1). To calculate GAB parameters, we rearranged eq 3 to the quadratic form[15] as follows

$$\frac{a_w}{M_{eq}} = \alpha \cdot a_w^2 + \beta . a_w + \gamma \tag{4}$$

where $\alpha = k_g \cdot (1 - C_g)/(M_0 \cdot C_g)$, $\beta = (C_g - 2)/(M_0 \cdot C_g)$, and $\gamma = 1/(M_0 \cdot C_g \cdot k_g)$.

Coefficients α, β, and γ were obtained by least-squares regression analysis of eq 4, and GAB parameters were obtained using the expressions

$$M_0 = \left(\frac{1}{\beta^2 - 4\alpha \cdot \gamma}\right)^{1/2}$$

$$k_g = -\frac{2 \cdot \alpha \cdot M_0}{\beta \cdot M_0 + 1}$$

$$C_g = \frac{1}{M_0 \cdot \gamma \cdot k_g}$$

2.3.4. Third Sorption Stage Model. The TSS formulation[13] predicts equilibrium moisture content as

$$M_{eq} = \frac{M_0 \times C_t \times H \times k_t \times a_w \times H'}{(1 - k_t a_w)(1 + (C_t H - 1) k_t a_w)} \tag{5}$$

H and H' are dimensionless intermediate functions defined by the equations

$$H = 1 + \frac{(1 - k_t)(k_t a_w)^h}{k_t (1 - a_w)} \tag{6}$$

$$H' = 1 + \frac{(H - 1)(1 - k_t a_w)}{H(1 - a_w)}[h + (1 - h) a_w] \tag{7}$$

where h is the third sorption stage constant and C_t and k_t are the TSS model constants relating to monolayer and multilayer sorption energies, respectively. When $h \rightarrow \infty$, H and H' approach unity, and eq 5 reverts to eq 3, the GAB model. To determine model parameters, we substituted eqs 6 and 7 into eq 5, and the new function was fitted to experimental data using OriginPro 8.

2.3.5. Parallel Exponential Kinetic Model. At each equilibration step, equilibrium moisture contents were simulated as a function of time using the empirical parallel exponential kinetic (PEK) model[16] according to the equation

$$M_t = M_{\infty 1}(1 - e^{-t/\tau_1}) + M_{\infty 2}(1 - e^{-t/\tau_2}) \tag{8}$$

M_t is moisture content at time t, whereas $M_{\infty 1}$ and $M_{\infty 2}$ are the moisture contents at the equilibrium state of two independent first-order kinetic processes, a "fast" process ($M_{\infty 1}$) and a "slow" process ($M_{\infty 2}$), representing simultaneous mass transfer in structurally different domains or sorption by energetically distinct mechanisms or sites. Parameters τ_1 and τ_2 are the characteristic times to obtain ~63% of $M_{\infty 1}$ and $M_{\infty 2}$, respectively, as explained by eq 9.

$$M_{\tau 1(\tau 2)} = M_{\infty 1(\infty 2)}(1 - e^{-1}) \approx M_{\infty 1(\infty 2)} \cdot 0.6321 \tag{9}$$

2.3.6. Young and Nelson Model. The Young and Nelson model[17] was used to simulate distribution of sorbed moisture, that is, tightly bound monomolecular layer, externally adsorbed moisture and internally absorbed water, in the first and second adsorption phases according to the function

$$M_s = A(\theta + \beta) + B\psi \tag{10}$$

where M_s is the adsorption equilibrium moisture content at any given a_w. Parameter θ is the surface fraction covered by monomolecular layer, ψ is the surface fraction covered by a layer of water that is two or more molecules thick, and β is the fraction of the total amount of adsorbed moisture in a multilayer, given by

$$\theta = \frac{a_w}{a_w + (1 - a_w)E} \tag{11}$$

$$\Psi = a_w \cdot \theta \tag{12}$$

$$\beta = -\frac{E \cdot a_w}{E - (E - 1)a_w} + \frac{E^2}{E - 1}\ln\frac{E - (E - 1)a_w}{E} - (E + 1)\ln(1 - a_w) \tag{13}$$

Constants A, B, and E are parameters characteristic of the material. Equations 11−13 were substituted into eq 10, and the resulting function was fitted to the experimental data using OriginPro 8. Arbitrary values of A, B, and E were initially assigned, and the function was allowed iterate to convergence with the lower value limit of each parameter preset at zero.

2.4. Statistical Analysis. Model goodness of fit was evaluated using coefficient of determination (R^2). As the measure of experimental dispersion of sorption data in the various models, normalized error function (NEF), which is root-mean-square error adjusted for monolayer capacity (M_0), was calculated as follows

$$\text{NEF}(\%) = \frac{100}{M_0} \times \left[\left(\sum_{i=1}^{n}(X_{\exp i} - X_{\text{calc}\, i})^2\right)/n\right]^{1/2} \tag{14}$$

$X_{\exp i}$ and $X_{\text{calc}\, i}$ are the experimental and fitted equilibrium moisture contents(g/100 g) at $a_{w\,i}$, respectively, and n is number of data points[18]

3. RESULTS AND DISCUSSION

3.1. Molecular Weight, Crystallinity, and Resistant Starch Contents of Samples. Molecular weight and polymer chain-length distribution influence the formation and stability of crystalline junctions that typify resistance to digestion and other technofunctional properties of recrystallized starch. The radius of gyration and molar mass distributions of the native and the acid-modified starch are presented in Figure 2. Acid-treated starch exhibited a smaller radius of gyration, which is consistent with reduction in chain length and branching pattern of the polymers. The number-average (M_n) and the weight-average (M_w) molar masses of NS were (1.5 and 5.3) × 10^5 g/mol, respectively, and decreased to (1.1 and 2.4) × 10^5 g/mol, respectively, representing a polydispersity index (M_w/M_n) drop from 3.53 to 2.18 in the acid-modified starch.

These molecular parameters, however, are lower than those reported for cassava and other starches in the literature. M_w of 5.7 × 10^7 to 4.2 × 10^8 g/mol was reported for cassava starch in comparison with, for instance, waxy corn and waxy rice starch, which have slightly higher molar masses.[19−21] The large disparities between our values and literature values could mainly be due to the analytical technique. According to Roger and coworkers,[22] asymmetric flow field flow fractionation might present difficulties in the characterization of the complete molar mass distribution of starch, particularly the high molar mass amylopectin fraction. Furthermore, the molecular parameters in our case are within those attributed to amylose.[22] Other factors such as differences in sample preparation protocols, data extrapolation method,[19,22] and variations in the ratio of amylose to amylopectin could also lead to disparate values. Nevertheless, a similar reduction in molar mass of starch due to acidification was reported.[20] NS granules are semicrystalline in nature, and during

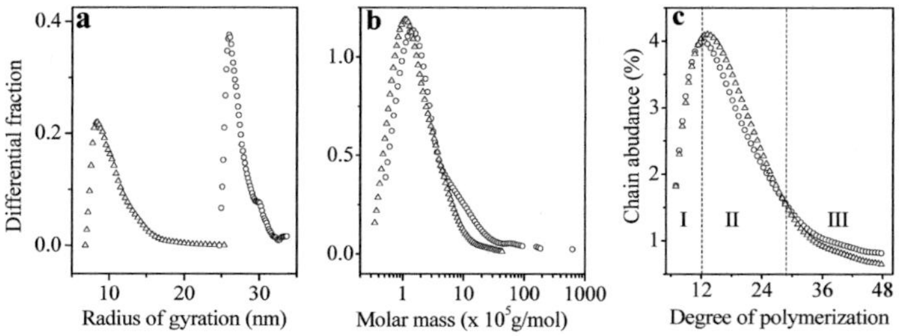

Figure 2. (a) Radius of gyration and (b) molar mass distributions of native (○) and acid-modified cassava starch (△) and (c) the polymer chain-length distributions of the debranched forms.

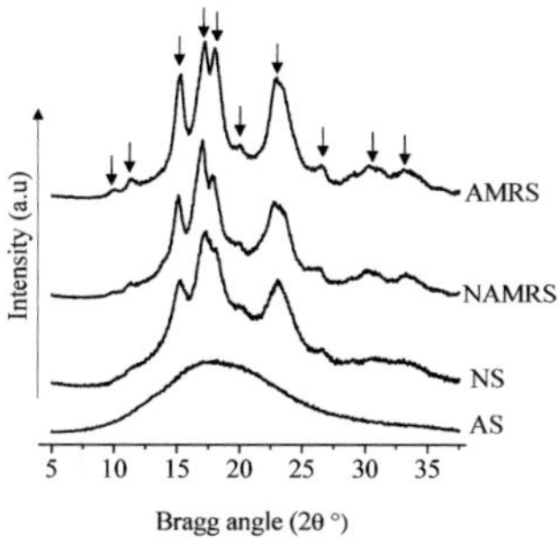

Figure 3. X-ray diffractograms of amorphous starch (AS), native starch (NS), non-acid-modified recrystallized starch (NAMRS), and acid-modified recrystallized starch (AMRS). Arrows indicate Bragg peaks used to distinguish and calculate starch crystallinity.

acid treatment, the amorphous domains, which comprise the long amylose chains and amylopectin branch points, are preferentially hydrolyzed.[23] Consequently, depending on starch type and conditions of hydrolysis, the starch residue after acid reaction comprises the more crystalline short-chain amylopectin domains and additional acid-resistant double-helical fragments of the amylose moiety.[24] These constituent short-chain polymers recrystallize more effectively into enzyme-resistant starch; a polymer chain length of 10−35 glucose units was shown to be favorable for starch recrystallization.[25] Polymer chain-length distributions of the native and acid-modified starches after debranching are presented in Figure 2c. This distribution is representative of cassava amylopectin side chains, whose apparent chain lengths range between 10 and 90 glucose units[26] with an average of 20 to 21.[27] Similar distributions, whereby the majority of these short polymers exhibit a chain length of ~12 glucose units, were also reported for debranched pea and banana starches.[28,29] Following acid treatment, the abundance of polymers comprising 12−28 glucose units (region II) increased, whereas the abundance of chains longer than 30 glucose units (region III) diminished, confirming that long polymers became hydrolyzed into shorter polymeric residues during acid reaction.

Figure 3 shows the X-ray diffractograms of the NS and the recrystallized products produced from the acid-modified and non-acid-modified starches, that is, acid-modified recrystallized starch (AMRS) and non-acid-modified recrystallized starch (NAMRS). A diffractogram of an amorphous starch (AS) sample is also included. NS exhibits strong reflections at 2θ = 15.2, 17.2, and 23.1°, and a shoulder is evident at 2θ = 18.1° typifying A-type crystalline packing, but the indistinct intensity at 2θ = 18.1° and the absence of the B-polymorph signature peak at 2θ = 5.5° suggest that the material subscribes to C-type crystallinity. Both A- and C-type packing patterns have been reported for native cassava starches, whereby C-type packing is a mixture of both A- and B-type patterns.[30] The recrystallized starches exhibit A-type crystalline packing structure, as demonstrated by the unambiguous doublet at 2θ = 17.1 and 18.1°, alongside other reflections at 2θ = 9.9, 11.2, 15.1, 22.9, 26.3, 30.3, and 33.1°.[31] Intensities of the Bragg peaks are lower in NS compared with NAMRS and AMRS, implying a lower crystallinity. The calculated degree of crystallinity of NS was 39.7% and corresponded well with literature values.[32] AMRS exhibited the highest crystallinity of 56.1% compared with 51.9% of NAMRS. These high crystallinities of AMRS and NAMRS are expected because rearrangement into double-helical order and recrystallization are enhanced when the starch polymers are linearized, for example, by selective enzymatic hydrolysis of the α-D-(1→6) branching points.[33] Recrystallization of AMRS, in particular, was favored by the absence of long polymer chains that cause impediment to intrapolymer hydrogen bonding due to their random conformation. The susceptibility of the starches to enzyme digestion was also examined. Contents of resistant starch were 44% in NS compared with 84 and 88% in NAMRS and AMRS, respectively. In NS, starch remaining after exposure to amylolytic enzymes accounts for resistant starch type II, which is due to compact structure of the ungelatinized granules. This structure limits accessibility of digestive enzymes. Strong intramolecular linkages between the linearized polymers were responsible for the high resistance to enzyme digestion of the recrystallized starches, meaning that accessibility to glycosidic bonds was more strongly hindered in these materials.

3.2. Sorption Isotherms and Hysteresis. Adsorption and desorption isotherms of the first sorption cycle are presented in Figure 4. All isotherms show an increase in equilibrium moisture content with increasing a_w and represent type II sorption according to Brunauer's classification.[34] Type II sorption isotherm is due to monolayer−multilayer sorption. A significant upswing at a_w close to 0.7 to 0.8 is evident from the isotherms. Such upward curvature may originate from liquefaction of vapor molecules in the void spaces of the materials.[13,35] This sorption behavior was

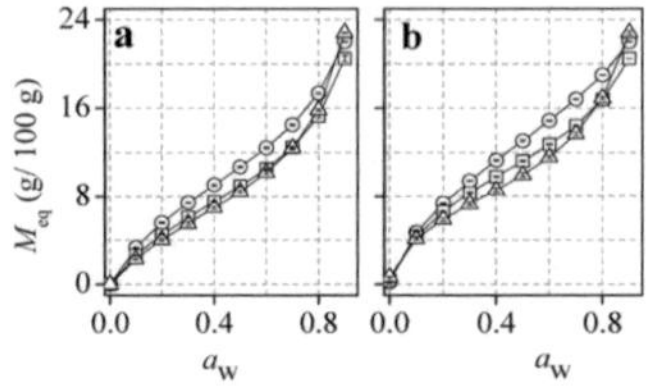

Figure 4. Isotherms of (a) adsorption and (b) desorption branches of the first sorption cycle for native starch (○), non-acid-modified recrystallized starch (□), and acid-modified recrystallized starch (△). Data are means of two determinations; error bars represent standard deviation.

also reported for cassava and other starches by other researchers.[32,36] Below $a_w \approx 0.8$, the recrystallized starches sorbed less water than NS. A similar observation was reported by Wronkowska and coworkers.[37] Sorption of water by starch is through interaction of water molecules with hydroxyl groups located mainly in amorphous regions and on the surface of crystallites; the crystalline regions typically resist moisture entry and have energetically less favorable sorption sites.[38] Therefore, sorption in the region below $a_w \approx 0.8$ corresponded well with the starch crystallinities. However, in the a_w region above 0.8, sorption patterns deviated from the observed crystallinity dependence; AMRS, despite having the highest crystallinity (56.1%), exhibited higher sorption capacity than NAMRS or NS, whose crystallinities were 51.9 and 39.7%, respectively. In general, moisture is capable of altering the structure of starch by either plasticization effects where polymers relax or unfold unraveling obscured polar sites or by crystallization effects whereby energetically favorable water binding sites become structurally concealed.[36]

Higher amounts of moisture were retained in the starches during desorption than were acquired during adsorption at any given a_w, and hysteresis of the respective starches is visualized in Figure 5. The degree of hysteresis (dh) is also plotted against a_w. Several factors including changes in polymer conformation, increase in the strength of adsorbent−adsorbate interaction (enthalpy effect), molecular ordering (entropy effect), or a combination of factors mediated by the entry of water into the polymeric structure have been suggested as possible causes for sorption hysteresis.[8,11] All three starches exhibited their highest dh in the low a_w region (Figure 5d), which is an indicator for intense enthalpy changes due to strong interactions between water and starch molecules.[39] The lower dh of NS as compared with NAMRS and AMRS in this low a_w region may, however, signify a relatively weaker specific interaction of its hydrophilic sites with zwater. NAMRS has higher dh than AMRS in the a_w region above 0.2, suggesting that the product is structurally less stable and the interaction with moisture opened up more sorption sites so that both sorption enthalpy and mass of strongly sorbed moisture increased by a larger magnitude in this product. NS and NAMRS, however, do not show a remarkably large difference in dh at $a_w > 0.7$, which might suggest that moisture detention mechanisms by the two materials in this region are relatively similar.

Following desorption, the starches were subjected to a second adsorption step with additional equilibrations at $a_w = 0.94$ and 0.98. The differences between equilibrium moisture contents of the second adsorption and the first adsorption in the range

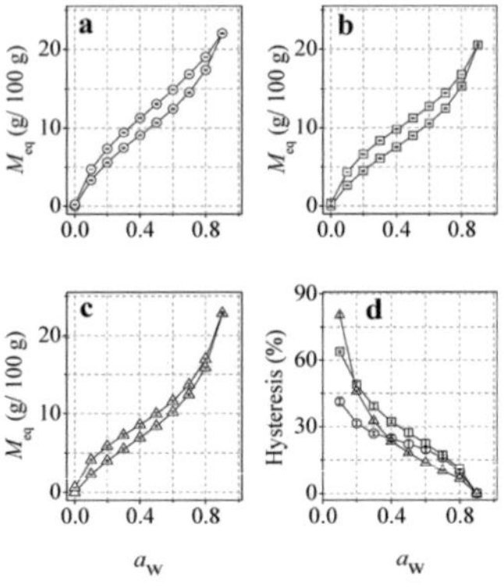

Figure 5. Sorption hysteresis for (a) native starch, (b) non-acid-modified recrystallized starch, and (c) acid-modified recrystallized starch and (d) the degree of hysteresis of the respective materials as a function of a_w. Data are means of two determinations; error bars represent standard deviation.

$0 < a_w < 0.90$ are plotted in Figure 6a. Clearly, although desorption was allowed to proceed to a steady state in all a_w levels, the initial "zero" moisture contents at $a_w = 0$ were not attained. Another significant observation is that the deviation of the equilibrium moisture contents of the second adsorption from the first ($M_{eq\,2} - M_{eq\,1}$) was highest at $a_w \approx 0.1$ to 0.2 and then decreased steadily, reaching negative values at $a_w > 0.7$ to 0.8. Besides, NS showed the least difference. A number of reasons account for these phenomena. In the wet state prior to desorption, all hydroxyl groups in the molecular structure of the starches are almost entirely occupied by the sorbed water molecules. During desorption, the hydroxyl groups are gradually freed, and these draw closer to form hydrogen bonds with each other. By this interaction, the void fraction of the matrices is probably irreversibly reduced so that moisture entrapment by capillary mechanisms is diminished. Furthermore, such intramolecular interaction is characteristic of crystallization, which reduces the void volume and elasticity of the matrices.[38] This argument, however, is less credible in the low a_w region, where the mobility of polymers is restricted. Some researchers, nevertheless, have argued that irreversible polymer shrinkage could cause kinetic barriers to moisture migration,[11] which might explain the "nonzero" moisture contents at $a_w = 0$ during desorption.

In Figure 6b, sorption of the starches in the high a_w region ($0.6 \leq a_w \leq 0.98$) during the second adsorption is illustrated. Evidently, at $a_w > 0.8$, NAMRS reached equilibrium at lower water content than AMRS, whereas at $a_w > 0.9$ and $a_w > 0.95$, NS reached equilibrium at lower water content than AMRS and NAMRS, respectively. The apparent isotherm upswing patterns are also of interest. One reason for the exponential increase in M_{eq} at high a_w is the formation of discrete liquid bridges between water molecules at the points where adjacent particles make contact, that is, capillary condensation.[13,35] This explanation has two implications. First, the a_w at which condensation begins to occur is related to the size of interparticulate spaces in packed beds; a lower onset a_w might suggest denser packing.[40] Second, decreased surface potential or incomplete wetting effects on the walls of

dx.doi.org/10.1021/bm101321q |Biomacromolecules 2011, 12, 660–671

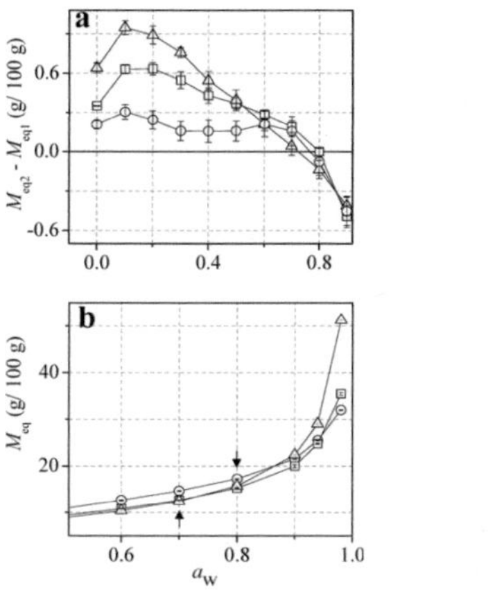

Figure 6. Deviation of second-adsorption equilibrium moisture contents ($M_{eq\,2}$) from first-adsorption equilibrium moisture contents ($M_{eq\,1}$) in the range $0 < a_w < 0.90$ (a) and isotherms of the second adsorption in the high vapor pressure region (b) for native starch (○), non-acid-modified recrystallized starch (□), and acid-modified recrystallized starch (△). Arrows in part b indicate onset points for isotherm upward inflection: AMRS arrow at $a_w = 0.7$; NS and NAMRS arrow at $a_w = 0.8$.

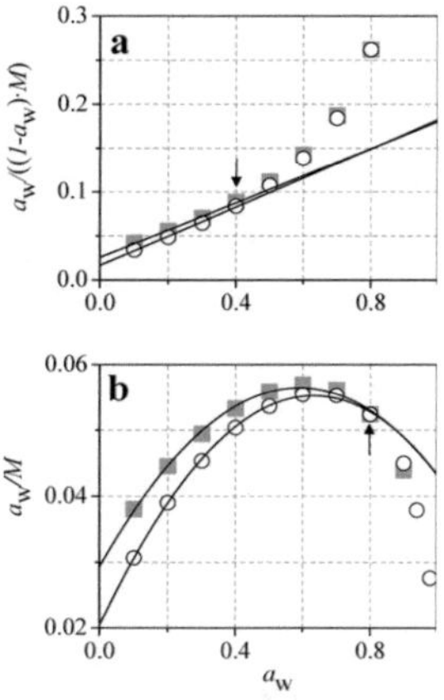

Figure 7. (a) Linear form of BET model and (b) parabolic form of GAB model fitted to the first (○) and second (■) adsorption isotherms of non-acid-modified recrystallized starch. Arrows show upper fit limits.

neighboring particles might enhance attraction forces between water vapor molecules, causing condensation to occur at a low relative vapor pressure.[41] In high a_w, starch molecules may also behave like Gaussian chains.[38] The hydroxyl groups along the chains participate in inter- and intramolecular hydrogen bonding that might give the starch a gel-like structure comprising networks of crystalline junctions and other loose entanglements. Occasionally, because of the network finite size, loose chains, or chain ends, part of the starch enters solution with adsorbed water.[38] The formation of such heterogeneous starch−water networks with different combinations of the gel and solution phases may have altered the chemical potential of moisture within the starches in different ways. A higher proportion of the ARMS was probably stabilized in the solution phase owing to the relatively smaller polymer size, whereas the branched nature and larger polymer sizes of NS and NAMRS supported a gel-like structure. Furthermore, because of the different structural states, the starches probably also formed different modified matrices characterized by pores of varying shapes, curvatures, and sizes, which might have significantly influenced the configuration of pore water and hence the amount that could be held mechanically by capillary forces.[40] Although it is difficult to state the real point of condensation onset, the isotherm upward inflection point is an indicator for liquefaction of vapor molecules in a material's interstitial spaces.[13,35] AMRS seemingly exhibited lower condensation onset vapor pressure among the starches as compared with NAMRS and NS because of a combination of these factors.

3.3. BET, GAB, and TSS Model Parameters. BET and GAB models were fitted to all isotherms of the first and second sorptions, whereas the TSS model, which is more relevant in describing sorption in the a_w region beyond 0.85 to 0.90,[35] was only fitted to the second adsorption isotherms. Because of the associated assumptions, the BET and GAB equations show limits in the respective ranges of applicability, which also differ depending on the adsorbent. For this reason, the appropriate fit limits were first identified using the inverse plot criteria described by Timmermann.[18] Figure 7 gives examples of experimental data fitted to linear (BET) and quadratic (GAB) forms. The respective model limits are indicated by arrows. In all cases, BET and GAB parameters were calculated on the basis of these limits, that is, $0.1 < a_w < 0.4$ and $0.1 < a_w < 0.8$, respectively. Fitting of experimental data by the models is illustrated in Figure 8. The BET relation overestimated the sorption in a_w region above 0.4, whereas the GAB model underestimated the sorption when a_w exceeded 0.8. These discrepancies relate to the specific model assumptions. The BET equation presupposes that the energy state of adsorbate molecules in the second and higher layers is the same as that of the pure liquid adsorbate. The GAB model improves the BET equation by introducing a differentiated second sorption stage based on the premise that the binding energy of some layers is intermediate between that of the monolayer and the pure liquid water but disregards the energy state of liquid-like water sorbed at high a_w.[18,35]

Table 1 presents BET and GAB parameters for the adsorption and desorption branches of the first sorption cycle, together with the respective coefficients of determination (R^2) and normalized error function (NEF). The R^2 values show that the models adequately represent the experimental data in the respective ranges of applicability. NEF was used as a more precise measure of the extent to which model values deviated from experimental data in the distinct models. NEF values <5% indicate a good representation.[18] It is seen, however, that GAB NEF values are lower than those of BET regression. The GAB fit was therefore better considering that the model covers a broader data range and so GAB parameters have been referred to in the subsequent discussion. Theoretically, in the GAB model (eq 3), C_g and k_g relate to the net heats of sorption of the monolayer and multilayer, respectively, and are described by the expressions: $C_g = C$

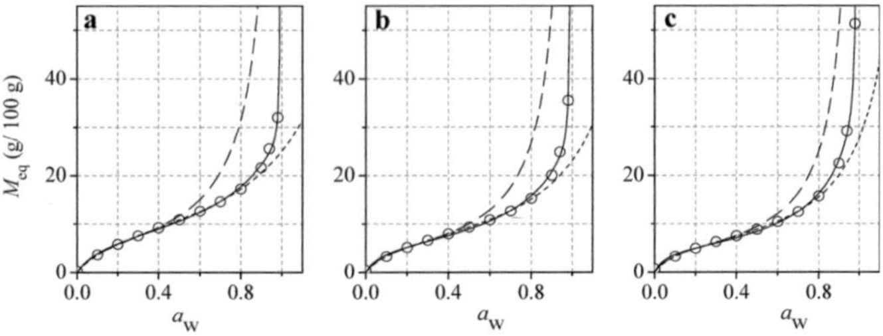

Figure 8. Experimental data (O) of the second adsorption modeled with BET (long dash), GAB (short dash), and TSS (solid line) functions for (a) native starch, (b) non-acid-modified recrystallized starch, and (c) acid-modified recrystallized starch.

Table 1. BET and GAB Coefficients and Measures of Fit for First Sorption Cycle

model[a]	adsorption			desorption		
	NS	NAMRS	AMRS	NS	NAMRS	AMRS
BET						
M_0 (g/100 g)	6.56 ± 0.24	5.54 ± 0.18	5.28 ± 0.15	7.76 ± 0.23	6.61 ± 0.23	5.61 ± 0.13
C_b (−)	7.99 ± 1.08	7.01 ± 0.77	5.94 ± 0.50	11.47 ± 1.68	14.24 ± 2.99	18.88 ± 3.40
R^2 (−)	0.991	0.993	0.994	0.995	0.993	0.997
NEF (%)	2.08	1.73	1.34	2.03	2.66	1.95
GAB						
M_0 (g/100 g)	9.21 ± 0.22	7.53 ± 0.21	6.68 ± 0.18	11.75 ± 0.20	9.25 ± 0.16	6.79 ± 0.09
C_g (−)	7.34 ± 0.21	6.52 ± 0.22	5.61 ± 0.20	9.95 ± 0.20	12.08 ± 0.36	15.41 ± 0.67
k_g (−)	0.654 ± 0.013	0.694 ± 0.014	0.770 ± 0.013	0.563 ± 0.010	0.611 ± 0.010	0.764 ± 0.008
R^2 (−)	0.997	0.993	0.987	0.999	0.999	0.999
NEF (%)	0.87	1.19	1.53	0.40	0.72	1.08

[a] Model ranges are BET: $0.1 \leq a_w \leq 0.4$; GAB: $0.1 \leq a_w \leq 0.8$. NS: native starch; NAMRS: non-acid-modified recrystallized starch; AMRS: acid-modified recrystallized starch

$\exp(H_0 - H_m)/RT$ and $k_g = k\exp(H_L - H_m)/RT$, where C and k are constants, H_0 is molar heat of sorption in the monolayer, H_m is molar heat of sorption in multilayers, H_L is molar heat of condensation of bulk water, R is the universal gas constant, and T is the absolute temperature.[11]

Monolayer capacity (M_0) for NS (9.21 g/100 g) was higher than that reported for native cassava starch by other authors (5.8 − 8.30 g/100 g) under similar temperature conditions.[42,43] Several factors such as the origin, ratio of amylose to amylopectin, surface properties of the granules (e.g., surface pores, surface proteins, and lipids), and the starch isolation procedures might influence moisture sorption by NS. Significantly disparate sorption values could also arise from the model fitting procedure, considering that practical model limits exist, as already pointed out. M_0 decreased in the order NS > NAMRS > AMRS, indicating a decreasing number or accessibility of adsorption sites. The monolayer adsorption enthalpy constants C_g decreased in the same order, suggesting that the driving force for monolayer binding also diminished. As expected, desorption M_0 values are generally higher than the adsorption M_0 and suggest that stronger binding forces increased the residence of water molecules on the starch surfaces. This is corroborated by the significantly higher C_g values predicted. Higher heats of desorption as compared with adsorption may be due to structural modifications that take place on the adsorbents surfaces during desorption, modifying the overall energy of binding of the adsorbate through cooperative binding or entrapment effects.[44]

Desorption C_g were higher by factors of 1.35, 1.85, and 2.75 in NS, NAMRS, and AMRS respectively, and the apparent values, unlike in adsorption, were inversely related to M_0. In terms of sorption phenomena, this observation implies that sorbed water changed the physical and probably also the chemical nature of the starch surfaces so that the binding stability of the monolayer as well as the number of hydrophilic sites increased. k_g, which relates to the chemical potential difference of the pure liquid state and the multilayer water molecules,[35] maintained a consistent trend (NS < NAMRS < AMRS) in both adsorption and desorption, meaning that the aforementioned moisture interaction effects were limited to the monolayer and did not affect the state of multilayer water. But the fact that k_g increased from NS to NAMRS to AMRS in both adsorption and desorption is an indicator that second stage sorption was less in NS than in NMRS and AMRS in that order, so that multilayer water was more structured toward the liquid-like state in AMRS than in NAMRS and NS.[35] Furthermore, increasing value of k_g also coincided with more pronounced upward curvature of isotherms in the high a_w region.

Model parameters of the second adsorption are summarized in Table 2. TSS parameters were derived from entire isotherms with additional equilibration data at 0.94 and 0.98 water activities, whereas the upper limits for BET and GAB models used in the

Table 2. BET, GAB and TSS Coefficients, and Measures of Fit for Second Adsorption

model[a]	NS	NAMRS	AMRS
BET			
M_0[b]	6.46 ± 0.18	5.49 ± 0.17	5.08 ± 0.12
C_b (−)	9.69 ± 1.17	11.05 ± 1.62	12.79 ± 1.71
R^2 (−)	0.995	0.994	0.996
NEF[c]	1.77	2.05	1.77
GAB			
M_0[b]	8.98 ± 0.06	7.20 ± 0.11	6.08 ± 0.11
C_g (−)	8.50 ± 0.08	9.72 ± 0.26	11.40 ± 0.51
k_g (−)	0.661 ± 0.004	0.701 ± 0.008	0.785 ± 0.009
R^2 (−)	0.999	0.999	0.997
NEF[c]	0.22	0.82	1.45
TSS			
M_0[b]	7.63 ± 0.44	6.06 ± 0.28	5.30 ± 0.16
C_t (−)	11.07 ± 2.48	13.71 ± 3.27	16.48 ± 3.43
k_t (−)	0.745 ± 0.018	0.789 ± 0.014	0.852 ± 0.007
h (−)	24.07 ± 2.06	24.98 ± 1.64	28.53 ± 1.19
R^2 (−)	0.997	0.998	0.999
NEF[c]	4.82	5.37	4.43

[a] Model ranges are BET: $0.1 \le a_w \le 0.4$; GAB: $0.1 \le a_w \le 0.8$; TSS: $0.1 \le a_w \le 0.98$. [b] Units are in g/100 g; [c] Expressed as a percentage. NS: native starch; NAMRS: non-acid-modified recrystallized starch; AMRS: acid-modified recrystallized starch.

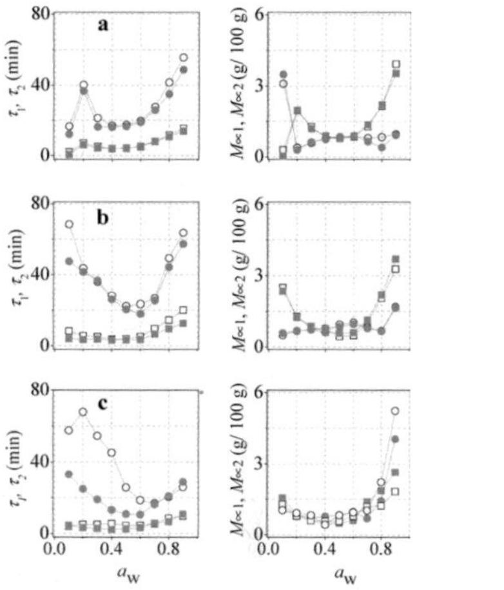

Figure 9. Characteristic times (left column) and mass parameters (right column) of first adsorption "fast" process (□), first adsorption "slow" process (○), second adsorption "fast" process (■), and second adsorption "slow" process (●) for (a) native starch, (b) non-acid-modified recrystallized starch, and (c) acid-modified recrystallized starch.

first sorption cycle (Figure 7) were retained. In all starches, second adsorption M_0 was slightly lower than that in the first adsorption. This is because sorption centers may have become structurally concealed, or some hydroxyl groups, freed in the preceding desorption, might have become unavailable for water binding because they participate in intramolecular bridging of the polymer chains. The energy parameters in the second adsorption were, however, higher as compared with those of the first adsorption. C_g, in particular, increased by factors of 1.2, 1.5, and 2.1 in NS, NAMRS, and AMRS, respectively, and was inversely related to M_0, unlike in the first adsorption. Relatively high monolayer adsorption energy may indicate the formation of an oriented chemisorbed layer of water molecules.[45] In the TSS model, the parameter h is of particular interest because it reflects the number of sorption layers that constitute the first and second sorption stages. The latter, which is multilayer sorption, proceeds to a limited number of layers, beyond which the third sorption stage begins when some of the sorbed water molecules attain liquid-like configuration at higher values of h.[35] h was not significantly different in NS and NAMRS but was remarkably higher in AMRS, indicating that adsorbed moisture was more liquid-like in this material.

3.4. PEK Modeling. A main advantage of a DVS analyzer is the ability to record sample mass changes in real time (Figure 1). Using Kohler's PEK model,[16] the kinetics of moisture sorption were simulated as a function of time. The "fast" process is related to adsorption in amorphous sites, binding on external surfaces, and direct binding on hydroxyl group. The "slow" process, conversely, is attributed to sorption in less accessible sites.[46] Figure 9 shows the characteristic times (τ_1, τ_2) and the characteristic masses ($M_{\infty 1}$, $M_{\infty 2}$) of the two processes during adsorption. For purposes of interpretation, characteristic times are inversely related to process rates in the distinct sites, whereas mass parameters relate to the respective sorption capacities. With increasing or decreasing a_w, the mass parameters represent the incremental adsorption or decremental desorption on each discrete a_w level.[16,46] As evident from Figure 9, both processes begin at a slow pace in the low a_w region. This can be attributed to adsorption kinetics, where vapor molecules have to diffuse from the bulk and appropriately orient themselves on the material's surface. As the a_w increases, equilibration rates also increase, reaching a maximum at a_w of ~0.6 due to a concurrent increase in spreading pressure, that is, surface potential,[47] and increasing plasticization effects. Thereafter, the processes proceed at a steadily decreasing rate (i.e., τ_1 and τ_2 increase), probably because the diffusion coefficient diminishes as the mechanism changes from vapor to liquid diffusion,[48] and sorption enthalpy declines as more water is adsorbed.[49] For NS, however, both adsorption processes initially occur quite rapidly; then, the rates slow down (kink at $a_w = 0.2$) before increasing again. The characteristic masses ($M_{\infty 1}$, $M_{\infty 2}$) show that this quick adsorption is dominated by the "slow" process which apparently is quite rapid compared with NAMRS and AMRS. A rapid initial adsorption is consistent with high affinity for moisture because of the low crystallinity. The kink at $a_w = 0.2$, however, may be explained in thermodynamic terms. Because of high moisture affinity, vapor molecules are rapidly drawn into sorption sites, causing the amorphous matrix to contract so that subsequent adsorption takes place at a slightly retarded pace as molecular attraction lessens and the matrix relaxes.[15] In addition, large amounts of heat evolved when water molecules vigorously adsorb might slow down the adsorption rate.[50]

PEK parameters of the second adsorption followed similar profiles as those of the first adsorption, particularly for NS and NAMRS (Figure 9). The values for τ_2 are lower, and remarkably low values are evident in AMRS in the a_w region below 0.7, meaning that equilibration occurred faster during the second adsorption. Regarding the mass parameters, it would appear that

dx.doi.org/10.1021/bm101321q | Biomacromolecules 2011, 12, 660–671

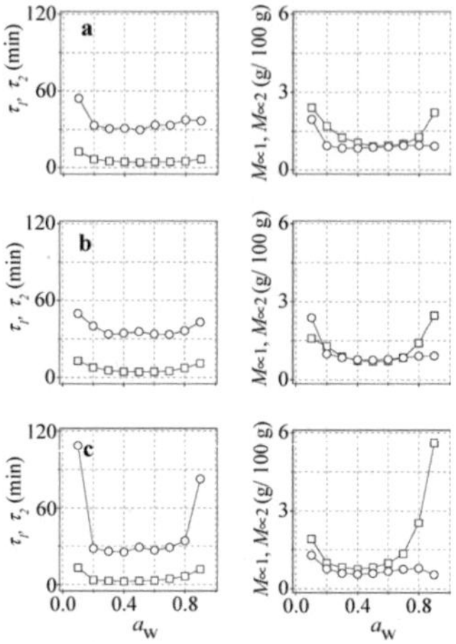

Figure 10. Characteristic times (left column) and mass parameters (right column) of "fast" (□) and "slow" (○) desorption processes of the first sorption cycle for (a) native starch, (b) non-acid-modified recrystallized starch, and (c) acid-modified recrystallized starch.

NS and NAMRS adsorbed more water by the "fast" process at the low and high a_w, whereas the two processes adsorb nearly equivalent amounts in the intermediate vapor pressure region. For AMRS, both the "slow" and "fast" processes adsorb equivalent amounts up to $a_w \approx 0.7$; thereafter, the "slow" process, which is evidently rapid as compared with NS and NAMRS, dominates. This means that the mechanism of adsorption in AMRS was remarkably different, particularly in the capillary condensation region.

PEK parameters relating to desorption in the first sorption cycle are presented in Figure 10. From the parameters τ_1 and τ_2, desorption begins at a slow rate in the high a_w. For AMRS, in particular, the rate of the "slow" process is distinctively low at the extreme water activities, although the amounts of water desorbed by this process are relatively low as compared with NS and NAMRS throughout the entire a_w range. This suggests that strong inter- and intramolecular interactions might have retarded vaporization at the capillary as well as diffusivity at the monolayer levels. The "fast" process dominates desorption, particularly in the capillary region, as seen from the parameters $M_{\infty 1}$ and $M_{\infty 2}$ in all of the starches. This is expected because when the vapor pressure of the surrounding is lowered, less strongly held water molecules at the surface are removed first before diffusional forces can begin to pull moisture out of the interior of the material. The difference between $M_{\infty 1}$ and $M_{\infty 2}$, especially in the a_w region above 0.7, might thus be a measure of the amount of "free" water detained mechanically in the interstitial spaces of the respective matrices.

3.5. Young and Nelson Model. Moisture adsorbed on material surfaces is subject to both binding and diffusional forces, the latter tending to cause internal absorption. We examined distribution of water adsorbed by the starches by fitting the experimental isotherms to the theoretical relationships of Young and Nelson.[17] In eq 10, $A = (\rho_w V_m)/W_m$ and $B = (\rho_w V_a)/W_m$, where V_m and V_a are the volumes (cubic centimeters) of adsorbed and absorbed water, ρ_w(g/cm^3) is the density of water, and W_m(g) is the mass of adsorbent. In eqs 11 and 13, $E = \exp[-(q_1 - q_L)/k_B T]$, where q_1 (J/mol) is the heat of adsorption of water to the polymer molecules, q_L (J/mol) is heat of condensation of water, k_B is Boltzmann's constant (1.38×10^{-23} J/K), and T (K) is temperature of the experimental conditions. Because the quantities q_1, V_m, and V_a are difficult to determine experimentally, regression techniques are often used to estimate the model coefficients.[8,49] The fitted values for A, B, and E, together with the measures of fit R^2 and NEF are given in Table 3. Parameter E was substituted into eqs 11 and 13 to obtain θ, β, and ψ at each a_w. The amount of moisture present as monomolecular water, externally adsorbed water, and internally absorbed water were computed as $A\theta$, $A(\theta + \beta)$, and $B\psi$, respectively. These are plotted in Figure 11.

A striking observation is that internal absorption occurred only in NS. The absorbed fraction increased linearly with increasing a_w, an indication that diffusional forces exceed binding forces. Monolayer coverage was fairly complete at $a_w \approx 0.6$, whereas the contribution of the externally adsorbed component to total sorption increased with increasing a_w. Regarding the second adsorption, internally absorbed moisture increased significantly, reaching 29.4% of the total adsorbed moisture at $a_w = 0.9$ as compared with 17.6% in the first adsorption. Concurrently, the fraction of externally adsorbed moisture increased, and a lower M_0 in the second adsorption is predicted, as also predicted by BET and GAB models, although with significantly lower values. The inability of adsorbed water to be absorbed in the internal matrices of NAMRS and AMRS has relevant consequences. First, the adsorbed moisture acquires enhanced capacity to form liquid bridges, and second, dissolution of small-size molecular components may intensify, compelled by increasing short-range dispersive or diminishing long-range structural forces as condensed moisture builds up on the materials' surfaces. In the event of the latter, solvated particles hydrate and supersaturate during desorption and are thus redeposited as an oriented film or form interparticulate solid bridges, which potentially alter the surface properties and pore structure of the matrices. As for NS, because of internal absorption, the amount of water condensing between particles was less. In other words, the polymer molecules in solution phase might have increased substantially in AMRS because of the combined effect of enhanced moisture condensation and small polymer size. These observations explain why the sorption of AMRS deviated by a relatively larger margin as compared with NAMRS and NS in the second adsorption (Figure 6) as well as the higher monolayer enthalpy constants (Table 2). The isotherm upswing phenomenon in the high a_w region (Figure 6) might be explained by solution phase formation. Furthermore, because dissolution proceeds slowly and is more relevant in the high a_w region, the remarkably high sorption by AMRS in the capillary condensation region is correctly modeled by the "slow" adsorption component of PEK simulation.

4. CONCLUSIONS

Model parameters of equilibrium moisture sorption of native and recrystallized cassava starches at 25 °C were compared. The NS sorbed more water at low-to-intermediate RH but was more stable to the restructuring effects of water. At the high RH, sorption by recrystallized starches exceeded that of NS because of capillary condensation and dissolution effects. By applying

Table 3. Young and Nelson Model Parameters, and Measures of Fit for First and Second Adsorptions

sample	model parameters			measures of fit	
	E (−)	A (−)	B (−)	R^2	NEF (%) [a]
1st adsorption					
NS	0.087 ± 0.009	5.31 ± 0.33	4.34 ± 1.20	0.999	2.90
NAMRS	0.166 ± 0.010	5.62 ± 0.03	0	0.999	2.74
AMRS	0.343 ± 0.064	5.60 ± 0.16	0	0.990	9.22
2nd adsorption [b]					
NS	0.053 ± 0.009	4.51 ± 0.35	7.06 ± 1.21	0.999	3.48
NAMRS	0.104 ± 0.007	5.77 ± 0.03	0	0.999	2.5
AMRS	0.186 ± 0.042	5.90 ± 0.13	0	0.988	10.65

[a] Calculated using $A\theta$ (g/100 g) at $a_w = 0.9$ as monolayer capacity. [b] Parameters are based on isotherm range $0.1 \le a_w \le 0.9$ for comparison with those of first adsorption. NS: native starch; NAMRS: non-acid-modified recrystallized starch; AMRS: acid-modified recrystallized starch.

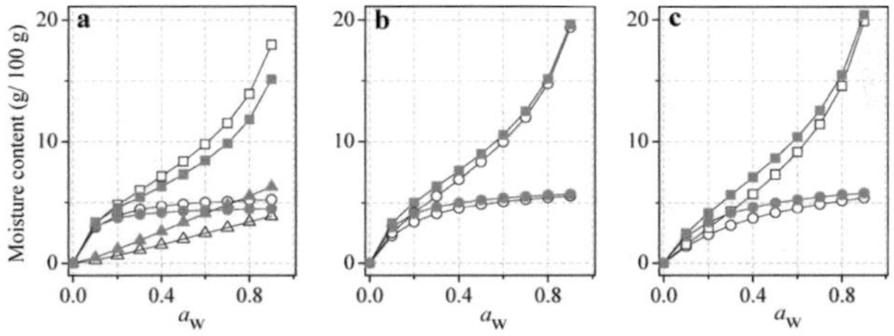

Figure 11. Monolayer (circle), externally adsorbed (square) and internally absorbed (triangle) moisture component isotherms for first (open symbols) and second (filled symbols) adsorptions of (a) native starch (b) nonacid modified recrystallized starch and (c) acid-modified recrystallized starch.

different models, it was possible to describe the physical states and distribution of adsorbed moisture. The GAB model was relevant in describing monolayer and multilayer sorption up to $a_w = 0.8$, whereas the TSS model enabled satisfactory isotherm fitting when a_w was extended to 0.98. Moisture was internally absorbed in the NS but not in the recrystallized starches. By not absorbing the adsorbed moisture, the recrystallized starches are likely to compete less for moisture in products where other macromolecular ingredients such as proteins, other carbohydrates, or even salts need the moisture for attainment of specific characteristics or in formulations where water in a less bound state is desirable for handling or end product qualities. This conclusion is particularly sensible where processing conditions are not too severe as to cause transitions like melting of the starches into forms that can compete more for moisture. It was possible to modify the water sorption of recrystallized starch by modifying the molecular structure of the starting material using mild-acid treatment. The lower moisture sorption of the smaller molecular weight (acid-modified) recrystallized starch makes it a better choice for low moisture products or products requiring moisture binding of the starch to be relatively low. However, dissolution effects driven by enhanced condensation and the small size of the constituent polymers might alter the hydrophilic character of fortified products upon exposure to high relative humidities. Failure of adsorbed moisture to diffuse into the internal matrices of the recrystallized starches also explains the high resistance of these starches to digestion because enzyme and starch might not approach each other over fairly sufficient surface.

■ AUTHOR INFORMATION

Corresponding Author

*Tel: +49-351-463-36943. Fax: +49-351-463-37126. E-mail: chrismutungi@yahoo.co.uk.

■ ACKNOWLEDGMENT

C.M. is grateful to Deutscher Akademischer Austauschdienst (DAAD) for the financial support. We thank Dr. Thomas Doert and Kathrin Trabs (Department of Chemistry and Food chemistry, Technische Universitaet Dresden) and Dr. Albena Lederer and Susanne Boye (Leibnitz Institute of Polymer Research, Dresden) for their assistance with X-ray diffraction and molecular weight analyses.

■ REFERENCES

(1) Sajilata, M. G.; Singhal, R. S.; Kulkarni, P. R. Resistant starch - a review. *Compr. Rev. Food Sci. Food Saf.* **2006**, *5*, 1–17.

(2) Nugent, A. P. Health properties of resistant starch. *BNF Nutr. Bull.* **2005**, *30*, 27–54.

(3) Sanz, T.; Salvador, A.; Fiszman, S. M. Evaluation of four types of resistant starch in muffin baking performance and relationship with batter rheology. *Eur. Food Res. Technol.* **2008**, *227*, 813–819.

(4) Duggan, E.; Noronha, N.; O'Riordan, E. D.; O'Sullivan, M. Effect of resistant starch on the water binding properties of imitation cheese. *J. Food Eng.* **2008**, *84*, 108–115.

(5) Ares, G.; Baixauli, R.; Sanz, T.; Varela, P.; Salvador, A. New functional fibre in milk puddings: effect on sensory properties and consumers' acceptability. *LWT—Food Sci. Technol.* **2009**, *42*, 710–716.

(6) Gelencser, T.; Gal, V.; Hodsagi, M.; Salgo, A. Evaluation of quality and digestibility characteristics of resistant starch-enriched pasta. *Food Bioprocess Technol.* **2008**, *1*, 171–179.

(7) Urso, M. E. D.; Lawrence, C. J.; Adams, M. J. Pendular, funicular, and capillary bridges: results for two dimensions. *J. Colloid Interface Sci.* **1999**, *220*, 42–56.

(8) Nokhodchi, A.; Ford, J. L.; Rubinstein, M. H. Studies on the interaction between water and (hydroxypropyl)methylcellulose. *J. Pharm. Sci.* **1997**, *86*, 608–615.

(9) Rizvi, S. S. H. Thermodynamic Properties of Food in Dehydration. In *Engineering Properties of Foods*, 2nd ed.; Rao, M. A., Rizvi, S. S. H., Eds.; Marcel-Dekker: New York, 1995; pp 223−309.

(10) Basu, S.; Shivhare, U. S.; Mujumdar, A. S. Models for sorption isotherms for foods: a review. *Drying Technol.* **2006**, *24*, 917–930.

(11) Al-Muhtaseb, A. H.; McMinn, W. A. M.; Magee, T. R. A. Moisture sorption isotherm characteristics of food products: a review. *Food Bioprod. Process.* **2002**, *80*, 118–128.

(12) van den Berg, C. Description of Water Activity of Foods for Engineering Purposes by Means of the GAB Model of Sorption. In *Engineering and Food*; McKenna, B. M., Ed.; Elsevier Applied Science: New York, 1984; Vol. 1, pp 311−321.

(13) Timmermann, E. O. A BET-like three sorption stage isotherm. *J. Chem. Soc., Faraday Trans. 1* **1989**, *85*, 1631–1645.

(14) Mutungi, C.; Onyango, C.; Rost, F.; Doert, T.; Jaros, D.; Rohm, H. Structural and physicochemical properties and in vitro digestibility of recrystallized linear α-D-(1 → 4) glucans derived from mild-acid-modified cassava starch. *Food Res. Int.* **2010**, *43*, 1144–1154.

(15) Figura, O. L.; Teixeira, A. A. *Food Physics: Physical Properties-Measurement and Applications.* Springer-Verlag: Berlin, 2007; pp 1−38.

(16) Kohler, R.; Duck, R.; Ausperger, B.; Alex, R. A numeric model for the kinetics of water vapor sorption on cellulosic reinforcement fibers. *Compos. Interfaces* **2003**, *10*, 255–276.

(17) Young, J. H.; Nelson, G. H. Theory of hysteresis between sorption and desorption isotherms in biological materials. *Trans. ASABE* **1967**, *10*, 260–263.

(18) Timmermann, E. O.; Chirife, J.; Iglesias, H. A. Water sorption isotherms of foods and foodstuffs: BET or GAB parameters? *J. Food Eng.* **2001**, *48*, 19–31.

(19) Yokoyama, W.; Renner-Nantz, J. J.; Shoemaker, C. F. Starch molecular mass and size by size-exclusion chromatography in DMSO-LiBr coupled with multiple angle laser light scattering. *Cereal Chem.* **1998**, *75*, 530–535.

(20) Bertolini, A. C.; Mestres, C.; Raffi, J.; Buleon, A.; Lerner, D.; Colonna, P. Photodegradation of cassava and corn starches. *J. Agric. Food Chem.* **2001**, *49*, 675–682.

(21) Tetchi, F. A.; Rolland-Sabate, A.; Amani, G. N.; Colonna, P. Molecular and physicochemical characterisation of starches from yam, cocoyam, cassava, sweet potato and ginger produced in the Ivory Coast. *J. Sci. Food Agric.* **2007**, *87*, 1906–1916.

(22) Roger, P.; Baud, B.; Colonna, P. Characterization of starch polysaccharides by flow field-flow fractionation-multi-angle laser light scattering-differential refractometer index. *J. Chromatogr. A* **2001**, *917*, 179–185.

(23) Hoover, R. Acid-treated starches. *Food Rev. Int.* **2000**, *16*, 369–392.

(24) Atichokudomchai, N.; Varavinit, S.; Chinachoti, P. A study of ordered structure in acid-modified tapioca starch by C-13 CP/MAS solid-state NMR. *Carbohydr. Polym.* **2004**, *58*, 383–389.

(25) Schmiedl, D.; Bauerlein, M.; Bengs, H.; Jacobasch, G. Production of heat-stable, butyrogenic resistant starch. *Carbohydr. Polym.* **2000**, *43*, 183–193.

(26) Ong, M. H.; Jumel, K.; Tokarczuk, P. F.; Blanshard, J. M. V.; Harding, S. E. Simultaneous determinations of the molecular-weight distributions of amyloses and the fine-structures of amylopectins of native starches. *Carbohydr. Res.* **1994**, *260*, 99–117.

(27) Charoenkul, N.; Uttapap, D.; Pathipanawat, W.; Takeda, Y. Molecular structure of starches from cassava varieties having different cooked root textures. *Starch/Staerke* **2006**, *58*, 443–452.

(28) Lehmann, U.; Jacobasch, G.; Schmiedl, D. Characterization of resistant starch type III from banana (*Musa acuminata*). *J. Agric. Food Chem.* **2002**, *50*, 5236–5240.

(29) Lehmann, U.; Rossler, C.; Schmiedl, D.; Jacobasch, G. Production and physicochemical characterization of resistant starch type III derived from pea starch. *Nahrung* **2003**, *47*, 60–63.

(30) Moorthy, S. N. Physicochemical and functional properties of tropical tuber starches: a review. *Starch/Staerke* **2002**, *54*, 559–592.

(31) Lopez-Rubio, A.; Flanagan, B. M.; Gilbert, E. P.; Gidley, M. J. A novel approach for calculating starch crystallinity and its correlation with double helix content: a combined XRD and NMR study. *Biopolymers* **2008**, *89*, 761–768.

(32) Perdomo, J.; Cova, A.; Sandoval, A. J.; Garcia, L.; Laredo, E.; Muller, A. J. Glass transition temperatures and water sorption isotherms of cassava starch. *Carbohydr. Polym.* **2009**, *76*, 305–313.

(33) Buleon, A.; Veronese, G.; Putaux, J. L. Self-association and crystallization of amylose. *Aust. J. Chem.* **2007**, *60*, 706–718.

(34) Brunauer, S.; Deming, L. S.; Deming, W. E.; Teller, E. On a theory of the van der Waals adsorption of gases. *J. Am. Chem. Soc.* **1940**, *62*, 1723–1732.

(35) Timmermann, E. O. Multilayer sorption parameters: BET or GAB values? *Colloids Surf., A* **2003**, *220*, 235–260.

(36) Al-Muhtaseb, A. H.; McMinn, W. A. M.; Magee, T. R. A. Water sorption isotherms of starch powders - Part 1: mathematical description of experimental data. *J. Food Eng.* **2004**, *61*, 297–307.

(37) Wronkowska, M.; Soral-Smietana, M.; Komorowska, E.; Lewandowicz, G. Y. Thermal and Water Sorption Isotherm Characteristics of Starch Preparations with High Content of Resistant Starch. In *Starch: From Polysaccharides to Granules, Simple and Mixture Gels*; Yuryev, V. P., Tomasik, P., Ruck, H., Eds.; Nova Science Publishers: New York, 2004; pp 105−118.

(38) Li, S.; Tang, J.; Chinachoti, P. Thermodynamics of starch-water systems: an analysis from solution-gel model on water sorption isotherms. *J. Polym. Sci., Part B: Polym. Phys.* **1996**, *34*, 2579–2589.

(39) Yang, W.; Sokhansanj, S.; Cenkowski, S.; Tang, J.; Wu, Y. A general model for sorption hysteresis in food materials. *J. Food Eng.* **1997**, *33*, 421–444.

(40) Bronlund, J.; Paterson, T. Moisture sorption isotherms for crystalline, amorphous and predominantly crystalline lactose powders. *Int. Dairy J.* **2004**, *14*, 247–254.

(41) Rouquerol, F.; Rouquerol, J.; Sing, K. *Adsorption by Powders and Porous Solids: Principles, Methodology and Applications*; Academic Press: San Diego, 1999.

(42) Sanni, L. O.; Atere, C.; Kuye, A. Moisture sorption isotherms of fufu and tapioca at different temperatures. *J. Food Eng.* **1997**, *34*, 203–212.

(43) Chatakanonda, P.; Dickinson, L. C.; Chinachoti, P. Mobility and distribution of water in cassava and potato starches by ^{1}H and ^{2}H NMR. *J. Agric. Food Chem.* **2003**, *51*, 7445–7449.

(44) McMinn, W. A. M.; Magee, T. R. A. Studies on the effect of temperature on the moisture sorption characteristics of potatoes. *J. Food Process Eng.* **1999**, *22*, 113–128.

(45) Czepirski, L.; Komorowska-Czepirska, E.; Szymonska, J. Fitting of different models for water vapour sorption on potato starch granules. *Appl. Surf. Sci.* **2002**, *196*, 150–153.

(46) Kohler, R.; Alex, R.; Duck, R.; Ausperger, B. A new kinetic model for water sorption isotherms of cellulosic materials. *Macromol. Symp.* **2006**, *244*, 89–96.

(47) Arslan, N.; Togrul, H. Modelling of water sorption isotherms of macaroni stored in a chamber under controlled humidity and thermodynamic approach. *J. Food Eng.* **2005**, *69*, 133–145.

(48) Yu, X.; Schmidt, A. R.; Bello-Perez, L. A.; Schmidt, S. J. Determination of the bulk moisture diffusion coefficient for corn starch using an automated water sorption instrument. *J. Agric. Food Chem.* **2008**, *56*, 50–58.

(49) Kachrimanis, K.; Noisternig, M. F.; Griesser, U. J.; Malamataris, S. Dynamic moisture sorption and desorption of standard and silicified microcrystalline cellulose. *Eur. J. Pharm. Biopharm.* **2006**, *64*, 307–315.
(50) Okubayashi, S.; Griesser, U. J.; Bechtold, T. A kinetic study of moisture sorption and desorption on lyocell fibers. *Carbohydr. Polym.* **2004**, *58*, 293–299.

Appendix 7

Mutungi, C., Passauer, L., Onyango, C., Jaros D., & Rohm, H. (2011b)

Debranched cassava starch crystallinity determination by Raman spectroscopy: correlation of features in Raman spectra with X-ray diffraction and ^{13}C CP/MAS NMR spectroscopy.

Carbohydrate Polymers, doi 10.1016/j.carbpol.2011.08.032.

Accepted Manuscript

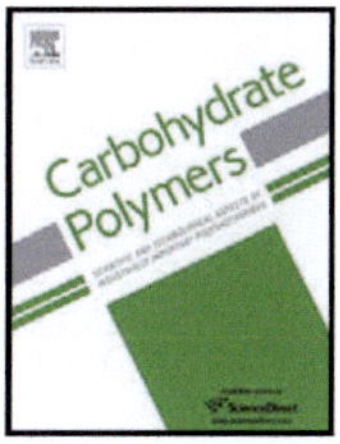

Title: Debranched cassava starch crystallinity determination by Raman spectroscopy: correlation of features in Raman spectra with X-ray diffraction and ^{13}C CP/MAS NMR spectroscopy.

Authors: Christopher Mutungi, Lars Passauer, Calvin Onyango, Doris Jaros, Harald Rohm

PII:	S0144-8617(11)00716-8
DOI:	doi:10.1016/j.carbpol.2011.08.032
Reference:	CARP 5886

To appear in:

Received date:	16-6-2011
Revised date:	9-8-2011
Accepted date:	11-8-2011

Please cite this article as: Mutungi, C., Passauer, L., Onyango, C., Jaros, D., & Rohm, H., Debranched cassava starch crystallinity determination by Raman spectroscopy: correlation of features in Raman spectra with X-ray diffraction and ^{13}C CP/MAS NMR spectroscopy., *Carbohydrate Polymers* (2010), doi:10.1016/j.carbpol.2011.08.032

Debranched cassava starch crystallinity determination by Raman spectroscopy: correlation of features in Raman spectra with X-ray diffraction and ^{13}C CP/MAS NMR spectroscopy.

Christopher Mutungi[a*], Lars Passauer[b], Calvin Onyango[c], Doris Jaros[a], Harald Rohm[a]

[a] Institute of Food Technology and Bioprocess Engineering, Technische Universitaet Dresden, D-01062 Dresden, Germany.

[b] Institute of Plant and Wood Chemistry, Technische Universitaet Dresden, D-01737 Tharandt, Germany.

[c] Food Technology Division, Kenya Industrial Research and Development Institute, P.O. Box 30650-00100 Nairobi, Kenya.

*Corresponding author: Tel: +49-351-463-36943, Fax: +49-351-463-37126, E-mail: chrismutungi@yahoo.co.uk (C. Mutungi)

Abstract

Because starch crystallinity influences the physical, mechanical, and technological aspects of numerous starch-based products during production and storage, rapid techniques for its assessment are vital. Samples of different levels of crystallinity were obtained by debranching gelatinized cassava starch, followed by subjection to various hydrothermal treatments. The recrystallized products were further subjected to partial hydrolysis with a mixture of α-amylase and glucoamylase prior to freeze-drying. Crystallinities were determined using X-ray diffraction (XRD) and ^{13}C CP/MAS NMR spectroscopy, and correlated with FT-Raman spectra features. XRD crystallinities ranged between 0 - 58 %, and agreed with crystalline-phase fractions (R^2=0.99) derived from the respective ^{13}C CP/MAS NMR spectra. A strong linear correlation was found between crystallinities and integrated areas of the skeletal mode Raman band at 480 cm^{-1} (R^2 = 0.99). With appropriate calibration, FT-Raman spectroscopy is a promising tool for rapid determination of starch crystallinity.

Keywords:

Starch; Crystallinity; X-ray diffraction; ^{13}C CP/MAS NMR; Raman spectroscopy

Introduction

Starch, the main storage polysaccharide in plants, is widely used in food, feed, pharmaceutical and other industrial applications. Amylose and amylopectin are the main constituent polymers. These are arranged systematically to form semi-crystalline granules that exhibit a degree of crystallinity of 15 - 45 % (Tester, Karkalas & Qi, 2004). Because of this semi-crystalline structure, native starch granules give X-ray diffraction patterns which have been used to categorize starches into A-, B- and C-polymorphs (Imberty, Buleon, Tran & Perez, 1991). Generally, cereal starches belong to A-polymorph whereas tuber and amylose-rich starches belong to B-polymorph. Legume and some tuber starches are categorized as C-polymorph (Tester et al., 2004).

The crystalline orientation of starch has been extensively studied (Tester et al., 2004). A- and B-polymorphs arise from left-handed parallel double-helical strands assembled in monoclinic and hexagonal lattice symmetries, respectively (Imberty et al., 1991). A higher density of packing of double helices, and fewer water molecules in the interstices of the A-polymorph as compared to the B-polymorph is the main structural difference between the two polymorphs. The C-polymorph is formed when both A- and B-polymorphic arrangements co-exist. Another crystalline arrangement, the V-form, is also categorized, but is based on single-helical glucopyranosyl chains (Buleon, Veronese & Putaux, 2007). These crystalline forms also appear when gelatinized starch recrystallizes under different conditions (Buleon et al., 2007; Mutungi, Rost, Onyango, Jaros & Rohm, 2009b).

Starch crystallinity affects the physical, mechanical, and technological properties of numerous starch-based products, and is therefore relevant for product development, quality and process control. In foods, loss of native crystallinity via gelatinization influences apparent viscosity, gelation and matrix forming characteristics, whereas reordering of the starch during processing or storage has impact product texture, stability, quality, digestibility and functionality (Sajilata, Singhal & Kulkarni, 2006). In pharmaceutical applications, starch may be used as an excipient. A certain degree of crystallinity is desired for the starch to maintain specific drug release and other functional properties (Pifferi, Santoro & Pedrani, 1999). In tablet formulation, for example, changes in crystallinity can occur during production processes such as drying, granulation, coating, compression, and milling because of exposure to mechanical stress, temperature changes, and different humidity levels. Amorphization or crystalline transformation can also occur during storage. The variations may modify properties such as compressibility or flowability or induce changes in solubility, dissolution or bioavailability of the active ingredient (de Veij, Vandenabeele, De Beer, Remonc & Moens, 2009). For these reasons techniques for the assessment of starch crystallinity are fundamentally relevant, especially if they offer possibilities rapid testing.

Infrared (IR) absorption spectroscopy has been used to follow starch crystallinity and molecular order because some spectral bands respond to these structural aspects (van Soest, Tournois, deWit & Vliegenthart, 1995). However, intensities of IR absorption bands are highly masked by intense absorption of water molecules, and are therefore poorly resolved. This effect renders quantitative interpretation of signals difficult. An alternative to IR-spectroscopy is Raman spectroscopy, which like IR-spectroscopy also probes vibrational transitions taking place in the ground electronic state of molecules. The method (Raman spectroscopy) measures the fraction of incident photon, usually monochromatic light from a visible or near infrared laser, which is scattered at an altered energy state due to vibrational transition energy of a molecule. The Raman signal is thus dependent on changes in polarizability of functional groups, whereas IR absorption signal is dependent on intrinsic dipole moment changes as the molecule vibrates (Ma & Phillips, 2002).

Raman spectroscopy exhibits minimal interference from moisture because water is only weakly polarizable. For this reason, numerous other bands typical of starch become apparent in the spectra, and can be evaluated (Thygesen, Lokke, Micklander & Engelsen, 2003). A number of studies have investigated the Raman signal of starch in response to molecular order loss during gelatinization (Kim, Yeh, Zhao & Wang, 1989; Schuster, Ehmoser, Gapes & Lendl, 2000) and molecular order gain during retrogradation (Bulkin, Kwak & Dea, 1987; Fechner, Wartewig, Kleinebudde & Neubert, 2005; Ma et al., 2002). Other investigators have also used Raman spectroscopy to detect differences in native (Dupuy & Laureyns, 2002; Santha, Sudha, Vijayakumari, Nayar & Moorthy, 1990; Wellner, Georget, Parker & Morris, 2011), chemically-modified (Dupuy et al., 2002; Passauer, Bender & Fischer, 2010; Phillips, Xing, Chong, Liu & Corke, 2000), and physically-modified starches (Kizil, Irudayaraj & Seetharaman, 2002). To the best of our knowledge, no discrete attempts have been made to correlate Raman signals with the long-range order of starch although spectra obtained for B- and V-amylose (Cael, Koenig & Blackwel.J, 1973) indicated that the technique is sensitive to conformational differences. Aim of this study was to investigate the extent to which Raman spectra relate to crystallinity of starch substrates as determined by wide angle X-ray diffraction, and solid-state ^{13}C cross-polarization and magic angle spinning nuclear magnetic resonance (^{13}C CP/MAS NMR) spectroscopy.

2 Materials and methods

2.1. Preparation of starch samples

Starch samples of different crystallinities were prepared from cassava starch. A 10 % w/w starch-water mixture was autoclaved at 121 °C for 15 minutes, and the cooled gel (50 °C) incubated with pullulanase (Sigma P2986, Sigma-Aldrich, Laborchemikalien GmbH, Steinheim, Germany) as described elsewhere (Mutungi, Onyango, Jaros, Henle & Rohm, 2009a). The debranched starch (DS) was freeze-dried, and the powder subjected to various recrystallization conditions: annealing (ANN), temperature-cycling (TC) or heat-moisture treatment (HMT). ANN-DS preparation involved autoclaving a 20 % w/w DS-water mixture at 121 °C for 15 min followed by incubation of the gel in a water-bath at 60 °C for 24 h. Preparation of TC-DS was done by subjecting a similar mixture to three autoclaving (121 °C for 2 h) and incubation cycles (60 °C for 1 h in the first two cycles, and 60 °C for 16 h in the third cycle). HMT-DS was prepared by moistening DS to 30 % w/w moisture content and autoclaving the wetted powder at 121 °C for 2 h. Portions of ANN-DS, TC-DS and HMT-DS were also subjected to a partial hydrolysis protocol, involving treatment with a solution of α-amylase (Sigma A3176) and amylogucosidase (Sigma A7095) at 37 °C for 16 h in a sodium acetate-hydrochloric acid buffer (pH 5.2), containing 0.02 % sodium azide (Mutungi et al., 2011). The partially digested samples were labelled dANN-DS, dTC-DS and dHMT-DS, respectively. All samples were freeze-dried and finely ground using motor and pestle.

2.2 Wide angle X-ray diffraction (XRD)

A Siemens D5000 (Bruker-AXS, Karlsruhe, Germany) X-ray diffractometer was used. Samples were scanned with Cu Kα radiation ($\lambda = 0.154$ nm), and reflections were detected in the angular range $2\theta = 5 - 37.5°$ with a step intervals of 0.01°. Scanning duration at each step was 6.5 s. The X-ray generator operating conditions were 40 kV and 25 mA. An amorphous sample prepared by autoclaving a 10 % (w/w) cassava starch-water suspension followed by freeze drying after storage at -82 °C for 20 min, was also scanned for background scattering estimation.

2.3 Nuclear magnetic resonance spectroscopy

Solid-state ^{13}C CP/MAS NMR measurements were performed on a Bruker Avance 300 spectrometer (Bruker BioSpin, Karlsruhe, Germany) operating at 75.48 MHz for the ^{13}C nucleus. A commercial double resonance 4 mm MAS NMR probe was used, and a ramped ^{1}H - ^{13}C cross-polarization applied. Sample spectra were accumulated with a spinning rate of 14 kHz at the magic angle (54.7°). The cross-polarisation contact time was 4 ms with a recycle delay of 2 s. Spectra were acquired with a line broadening of 10 Hz. At least 2000 scans were accumulated for each spectrum at room temperature. Spinal ^{1}H-decoupling was applied during signal acquisition to remove proton spin-spin interactions and improve spectra resolution. Spectra were referenced indirectly to a zero value for tetramethylsilane. Quantitative evaluation of spectra involved decomposition of each spectrum into the respective amorphous- and ordered-phase signatures by subtracting a linearly scaled spectrum of the amorphous starch sample so that the intensity of the difference (ordered-phase) sub-spectrum was zero at 84 ppm (Paris, Bizot, Emery, Buzare & Buleon, 1999). Scaling of the amorphous spectrum and spectral subtraction were done on Microsoft excel® solver. The ordered-phase sub-spectra were then fitted with deconvolution peaks, and integrated areas of peaks within the C1 atom signal were used in calculation of crystalline-phase contents. PeakFit® version 4.12 was used for spectra processing.

2.4 Fourier Transform-Raman spectroscopy (FT-Raman)

Raman spectra were recorded on a Bruker MultiRAM FT-Raman spectrometer (Bruker Optics, Ettlingen, Germany) equipped with an air-cooled neodymium-doped yttrium aluminium garnet (Nd:YAG) laser source operating at $\lambda_{Nd:YAG}$ = 1064 nm, and a liquid nitrogen-cooled germanium diode detector. Samples were placed on an aluminium holder and 200 scans were accumulated and co-added. The spectral resolution was 4 cm^{-1}. Evaluation of acquired spectra involved integrating band areas, band widths, and wavenumber at maximum height of bands in the region 3500 - 300 cm^{-1} after baseline correction and vector normalization against the C-O-C symmetric stretch mode at ~865 cm^{-1} using the manufacturer's software OPUS/QUANT version 6.

Results and discussion

3.1 XRD crystallinity

Aqueous starch re-associates by forming double-helical strands along the linear sections of constituent polymer chains. These strands subsequently aggregate to ordered domains of different morphologic orientation depending on recrystallization conditions (Buleon et al., 2007). One way to increase starch recrystallization is by debranching it with enzymes that hydrolyze α-(1 → 6) bonds in the amylopectin, and in some amylose chains (Mutungi et al., 2009a; Pohu, Planchot, Putaux, Colonna & Buleon, 2004). However, recrystallization of starch is a complex process involving conformational change, chain alignment, crystal packing, and phase propagation (Bulkin et al., 1987). Whereas debranching frees short linear α-(1 → 4) polymers that favor crystallization, the longer amylose chains associate much faster and form networks that impede propagation of substantially ordered crystallites. Enhancement of recrystallization is achieved by subjecting the debranched starch to hydrothermal treatments (Mutungi et al., 2009b). The effects of various hydrothermal conditions classified as annealing, temperature cycling and heat-moisture treatment were reported in an earlier investigation (Mutungi et al., 2009b); the different combinations of thermal and plasticization conditions induce chain ordering effects that result in products of different crystallinity and crystalline orientations.

Figure 1 shows X-ray diffractograms of the various starch preparations. Evolution patterns, and intensities of individual reflections particularly at $2\theta = \sim 10°$, 11°, 15°, 17°, 18°, 20°, 23°, 24°, 26°, 30°, and 33° indicate that the materials comprise mainly different contents of A-, and B-polymorphs, and small amounts of V-type structures. A-polymorph is frequently distinguished by strong reflections at $2\theta = 15.11°$, 17.14°, 18.14°, and 26.27°, with some additional reflections at $2\theta = 9.98°$, 11.19°, 22.93°, 23.68°, 30.30° and 33.08° (Lopez-Rubio, Flanagan, Gilbert & Gidley, 2008), whereas B-polymorph is recognized by typical reflections at $2\theta = 5.51°$, 14.60° and 16.85° with auxiliary peaks at

2θ = 10.01°, 11.02°, 13.85°, 22.30°, 23.71°, 26.16°, 30.61° and 33.84°. V-polymorph is identified by a reflection at 2θ = 19.8° and occasionally, additional reflections occur at 2θ = 7.4° and 13.0° (Lopez-Rubio et al., 2008).

HMT and dHMT subscribe more to A-type crystallinity especially because the respective diffractograms exhibit a clear A-polymorph signature doublet at 2θ = 17.1° and 18.1° in addition to strong reflections at 2θ =15.1°. The rest of the samples are varying mixtures of A- and B-polymorphs (C-polymorph) because the respective XRD patterns neither exhibit the A-polymorph signature doublet (2θ = 17.1° and 18.1°) nor the B-polymorph signature reflection, which is usually a strong peak at ~2θ = 5.5° (Cairns, Bogracheva, Ring, Hadley & Morris, 1997). The reflection at $2\theta \approx 20°$ is low, and indicates that only small amounts of V-type structures are regularly packed so as to diffract X-rays. Furthermore, some intensity at this angular position can also arise from A- or B-crystalline packing (Lopez-Rubio et al., 2008).

From the intensities of reflections (arrows Figure 1) the various samples exhibit different crystallinity levels. A common method used to quantify crystallinity of starch involves fitting a smooth curve under the main minima of diffractograms (Figure 2), as originally proposed by Hermans and Weindiger (Hermans & Weidinger, 1948) for cellulose. The method assumes a two-phase situation, whereby relatively perfect crystalline domains are interspersed in an amorphous background, so that crystallites give distinct reflections. However, ordered starch often exhibits imperfections in the form of chain foldings, entangled chain ends, or branched-chain segments that are constrained within crystallite assemblies (Lopez-Rubio et al., 2008). These non-crystalline interfaces cause fluctuations in the relative distances between corresponding atoms in the ideal lattice, and therefore some scattering ability of crystallites is lost or becomes diffuse. Furthermore, small crystallites that possess only a few lattice planes may not produce sufficiently detectable reflection intensities. This diffuse intensity is erroneously estimated together with the amorphous background when the two-phase approach is adopted (Gernat, Radosta, Anger & Damaschun, 1993; Lopez-Rubio et al., 2008).

To improve accuracy of crystallinity estimation, Lopez-Rubio et al., (2008) proposed a method for modeling the amorphous contribution by fitting a Gaussian profile to the experimental diffractogram. Whereas the approach could account for lattice imperfections, a baseline derived from an appropriate amorphous template is easier to fit, and also corrects for instrumental background scattering as well (Cairns et al., 1997). In Figure 2, profiles of diffractogram-minima baseline and amorphous-template baseline fittings are illustrated. The amorphous template is fitted under the sample diffractogram with respect to the two extreme ends, so that no negative intensities are generated on the difference sub-diffractogram upon subtraction. Clearly the amorphous scattering does not extend up to the base of diffraction peaks as projected by the diffractogram-minima baseline. Intensity difference between the diffractogram-minima baseline and the amorphous-template baseline (Figure 2b) may be taken as measure of diffuse scattering arising from imperfect and small crystallites. The crystallinity values calculated using the two approaches are presented in Table 1.

Crystallinity increased in the order DS < ANN-DS < TC-DS < HMT-DS in both approaches, and values obtained by the amorphous-template approach were substantially higher. However, for the partially hydrolyzed samples, that is, dANN-DS, dTC-DS, and dHMT-DS, derived from ANN-DS, TC-DS, and HMT-DS, respectively, the amorphous-template approach revealed an increase in crystallinity in each case (dANN-DS > ANN-DS; dTC-DS > TC-DS and dHMT-DS > HMT-DS), whereas the diffractogram-minima baseline approach gave lower crystallinity values (dANN-DS < ANN-DS; dTC-DS < TC-DS and dHMT-DS < HMT-DS). This disagreement is explained by the fact that diffractogram-minima baseline approach does not account for diffuse scattering, which might have increased in the partially hydrolyzed samples due to crystallite erosion and disentanglement of packed chains during enzyme action (Mutungi et al., 2011). Furthermore, a less rigid chain packing in the hydrolyzed samples is especially implied by the loss of reflection intensities at $2\theta = 15°$ and 18° (Figure 1). Strong reflections in these angular positions typify the more dense A-type crystalline structure.

3.2 Crystalline-phase by ^{13}C CP/MAS NMR

Starch polymers in crystalline and non-crystalline assemblies possess different α-(1 → 4) glycosidic bond torsion angles. This implies different local electron densities around carbon atoms in the different domains due to conformational effects (Paris et al., 1999). The different conformational populations can be probed using solid-state ^{13}C CP/MAS NMR because a straight forward correlation exists between chain conformation and the peak profile of the C1 atom resonance signal (Gidley & Bociek, 1988; Paris, Bizot, Emery, Buzare & Bulcon, 2001). Regularly arranged conformations of crystalline packed chains correspond to narrow resonance frequency C1 bands with multiplicities determined by space group equivalent symmetry classes, whereas chain conformations that are not crystalline become sufficiently distorted on an average so as not to show up at similar chemical shifts (Paris et al., 1999). By deconvolution of the C1 signal, chains that are packed into crystalline domains, and those that are not, can thus be conveniently delineated (Mutungi et al., 2011; Paris et al., 2001). Moreover, the intensity of the C4 atom signal relates directly to the content of the amorphous phase in a sample (Gidley et al., 1988).

^{13}C CP/MAS spectra of the different starches are shown in Figure 3. Spectral consistency was tested by evaluating the intensity ratio between the combined intensities of C2, C3, C4 and C5 resonance signals and the intensity of C1 resonance signal (Paris et al., 1999, Tan, Flanagan, Halley, Whittaker & Gidley, 2007). This ratio was 3.97 ± 0.12 for the different samples spectra, and the constancy provided confidence that the amorphous and crystalline regions behaved fairly similarly with respect to their solid state nature as probed by the NMR experiment. The intensity of C4 signal progressively decreased in the order AS > DS > ANN-DS > TC-DS > HMT-DS, signifying decreasing amount of the amorphous-phase, and was indistinct among the partially digested substrates (i.e. dANN-DS, dTC-DS and dHMT-DS). These findings are in agreement with the XRD crystallinities derived using the amorphous-template approach reported in Table 1, which depict increase in crystallinity following partial hydrolysis of the materials. Amorphous- and ordered-phases were quantified by decomposing NMR spectra based on the assumption that each individual spectrum is a linear combination of amorphous- and ordered-phase signatures (Paris et al., 1999). An example of spectrum decomposition and the resultant sub-spectra is illustrated in Figure 4. Amorphous-phase content was calculated from the C1 signal (90 - 107 ppm) as percent ratio of the difference in integrated intensity before and after spectrum decomposition relative to the integrated intensity before decomposition. The values are presented in Table 1. Amorphous-phase contents were not determined for dANN-DS, dTC-DS, dHMT-DS as the C4 signal was too weak in these samples (Figure 3). Lack of distinct signals at the C4 chemical shift region indicated that the amorphous fraction in the partially digested samples was hydrolyzed by amylolytic enzymes, resulting in increased molecular ordering of the remaining crystalline fraction (Mutungi et al., 2011).

In Figure 3, profiles of the deconvoluted C1 atom signals are also illustrated. The two sets of peak multiplicities, a triplet and a doublet, characteristic of A- and B-crystalline packing respectively, are evident. A triplet appears in A-type starch because double helices packed in the typical monoclinic lattice, depict three equivalent classes for the six monomers that constitute a single helical turn; the three classes correspond to three distinct torsions about the α - (1 $\rightarrow$ 4) linkage. A doublet, on the contrary, appears in B-crystalline packing for the reason that two equivalent classes of α - (1 $\rightarrow$ 4) bond torsions are exhibited among the hexagonally packed chain duplexes (Paris et al., 1999). The multiplicity sets produce narrow line contributions at approximately 99, 100 and 101 ppm (intensity ratio 1:1:1) and at about100 and 101 ppm (intensity ratio 1:1). For this reason, crystalline-phase was calculated from the relative intensity contribution of these lines in the C1 atom signal of the ordered-phase sub-spectra as illustrated in Figure 4c. Once a doublet or triplet was assigned, three other peaks representing contributions related to interfacial conformations were allocated at approximately 94, 97 and 103 ppm as discussed elsewhere (Mutungi et al., 2011; Paris et al., 2001). In comparison to Gaussian and Lorentzian functions, Voigt function provided the most reproducible fitting of peaks with approximately equal intensities for the triplet or doublet. The Voigt function combines both Gaussian and Lorentzian profiles and is defined by the equation:

$$y = f \cdot \frac{\lambda}{\pi} \int_{-\infty}^{\infty} \frac{\exp(-t^2)}{\lambda^2 + (\sigma - t)^2} dt$$

where $f = \frac{1}{w_G}(\ln 2)^{1/2}$; $\lambda = \frac{w_L}{w_G}(\ln 2)^{1/2}$; $\sigma = \frac{x - x_c}{w_G}(\ln 2)^{1/2}$; and w_G and w_L are full width at half maximum (FWHM) of the Gaussian and the Lorentzian profiles, respectively, x_c is the centre and t is integral parameter. Contributions by crystalline- and interfacial-phases in the various starches, expressed as percent fraction relative to the total integrated intensity of the C1 signal before spectral decomposition are given in Table 1. NMR crystalline-phase contents closely correspond to the XRD crystallinities derived by the amorphous-template baseline approach. The latter data (XRD crystallinities by amorphous-template fitting) were correlated with Raman spectra features, and only these have been referenced to in the subsequent discussion.

3.3 Correlation of crystallinity with features in Raman spectra

FT-Raman spectra of the various starch preparations are presented in Figure 5, whereas the assignments to bands are summarized in Table 2. The band at about 865 cm^{-1} was used as internal standard to normalize the spectra for sample size. This band is assigned to symmetric C(1)-O-C(5) stretching (breathing mode) of the α-D-glucose ring (Cael,

Koenig & Blackwell, 1975), whose intensity should not be influenced by the preparation of the starch samples. A number of band dissimilarities are discernible including splitting, sharpening, relative heights, and wavenumber at maximum heights (σ_{Hmax}). Overlaid profiles of some bands are shown in Figure 6a.

Splitting of the band at 1381 cm^{-1}, usually assigned to δ(C-OH) and δ(C-H) modes (Cael et al., 1973) generated a shoulder at 1403 cm^{-1} assigned to δ(C-H). The prominence of this shoulder increased with increasing starch crystallinity, implying crystallinity-dependent decoupling of the deformational modes (C-H and C-OH) at 1381 cm^{-1}. Similarly the shoulder at 1155 cm^{-1} associated with α-(C1-O-C4') asymmetric stretch (Sekkal, Dincq, Legrand & Huvenne, 1995), as well as the shoulder at 1110 cm^{-1} related to C-C, C-H and C-O modes became more prominent due to splitting of the complex band in the 1175-1095 cm^{-1} region.

Band sharpening is evident for the bands at 1264 cm^{-1}, 943 cm^{-1}, 865 cm^{-1} and 480 cm^{-1}, which are related to $\delta(CH_2)$, υ_s(C1-O-C4'), υ_s(C1-O-C5), and skeletal (C-C-O) modes, respectively (Cael et al., 1973). The respective σ_{Hmax} of these bands also shifted markedly with increasing crystallinity. The σ_{Hmax} of the (C-C-O) band at 480 cm^{-1} shifted to a lower value ($r = 0.95$; $p < 0.001$; n = 8). A shift of this band towards a higher wavenumber was previously reported during starch gelatinization (Fechner et al., 2005). Likewise, the σ_{Hmax} of the bands at 1264 cm^{-1}, 865 cm^{-1} and 943 cm^{-1} shifted to higher values ($r = 0.98$, 0.96, 0.96, respectively). The bands at 1264 cm^{-1} and 943 cm^{-1} were reported to shift to lower wavenumbers following transformation of amylose from V- to B-form in a mechanism involving dissociation of the CH_2OH mediated intra-chain bridges due to uptake of water (Cael et al., 1973). Such a mechanism also occurs during starch gelatinization (Schuster et al., 2000). From the XRD reflections of the starches (Figure 1), contribution by V-structures is not the cause of the shifts of these bands to higher wavenumbers in our case. Instead, an increase in crystallinity arising from helicalization and stabilization of the double helical polymers via intra-chain hydrogen bonding is a more plausible explanation.

Band widths (full width at half maximum height, FWHM) may also provide a measure of the structural variation between samples, whereby, band narrowing is indicator for a narrower distribution of bond energies in the more ordered specimens (Bulkin et al., 1987). This parameter was correlated with crystallinity in the relatively strong bands at 2910 cm^{-1}, 1264 cm^{-1}, 943 cm^{-1}, 865 cm^{-1}, and 480 cm^{-1}. With increasing starch crystallinity, FWHM declined in these bands. For $n = 8$, the correlation coefficients (r) were 0.93 ($p < 0.001$), 0.82 ($p < 0.05$), 0.89 ($p < 0.01$), 0.89 ($p < 0.01$) and 0.98 ($p < 0.001$), respectively. Bulkin et al., (1987) and Fechner et al., (2005) observed similar narrowing of the bands at 2910 cm^{-1} and 480 cm^{-1} in starch gels undergoing retrogradation. From our results, FWMH is most linearly responsive to crystallinity in the skeletal mode band at 480 cm^{-1}. Elsewhere, the 480 cm^{-1} band was reported to become narrower in the spectra of malto-oligosaccharides as the chain length increased to ~60 glycosyl units beyond which no further change occurred (Bulkin et al., 1987). This chain-length effect was seemingly insignificant with our samples.

The υ(C-H) band at 2910 cm^{-1} also gave a relatively strong correlation between FWHM and crystallinity ($r = 0.93$; $p < 0.001$; n = 8), an indicator that C-H stretching modes are fairly sensitive to conformational order and packing of polymer chains (Bulkin et al., 1987). However, the correlation coefficient was slightly lower as compared to that of the (C-C-O) band at 480 cm^{-1} ($r = 0.98$; $p < 0.001$; n = 8), probably because the υ(C-H) band is a collection of signals arising from symmetric and asymmetric stretching modes of the methylene group (Cael et al., 1973), and possibly also from contributions of Fermi resonance between the overtone of CH_2 deformation and stretching fundamentals (Bulkin et al., 1987).

The moderately strong correlations between crystallinity and FWHM of the υ_s(C1-O-C4') and υ_s(C1-O-C5) bands at 943 cm^{-1} and 865 cm^{-1}, respectively ($r = 0.89$; $p < 0.01$ in both cases) may be explained in terms of double helix formation. Orientation of the starch polymers into helical forms coincides with a chain twist about α-(1 → 4) glycosidic bonds, and the characteristic torsion angles impose a strain-related loss of the stretching symmetry of C1-O-C4' atoms responsible for the band at ~ 943 cm^{-1}(Cael et al., 1973; Sekkal et al., 1995). Similarly, increasing chain helicalization progressively alters the orientation of glycosyl units and therefore the symmetry of C(1)-O-C(5) stretch, as well as the coupled deformations of C(1)-H, C(5)-H and C(6)-H_2 groups (Cael et al., 1975). The relatively weaker correlation between FWHM and crystallinity at the 1264 cm^{-1} band ($r = 0.82$; $p < 0.05$; n = 8) is probably because vibrations in this region also involve C-OH related modes, meaning that the hydration degree of specimens might have confounding effects.

An advantage of using FWHM for spectral evaluation is that FWHM is a true spectroscopic parameter that is not proportional to concentration, and is less affected by the amount of sample and other experimental variables. However, weak bands that might provide precise relations because they arise from more independent modes are difficult to accurately appraise using this parameter. To incorporate these weak modes in our spectral evaluation, band areas were integrated. Correlations of areas of selected bands and starch crystallinities are shown in Figure 6b.

Areas of C-C-O band at 480 cm^{-1} gave the best correlation with starch crystallinities (r = 0.99; $p < 0.001$; n = 8). This band, which is assigned to symmetric vibrations of the glucopyranose ring (Cael et al., 1973), is characteristically strong in starch as compared to other polysaccharides (de Veij et al., 2009). Areas of some other bands also correlated well with starch crystallinity. Schuster et al., (2000) have reported that Raman intensities of the majority of modes including (C-C-O) at 480 cm^{-1} and 530 cm^{-1}; δ(C-OH) and δ(C-H) at 910 cm^{-1}; υ_s(C1-O-C4') at 943 cm^{-1}; υ(C-OH) and δ(C-OH) at 1053 cm^{-1}; δ (C-OH) at 1084 cm^{-1}; υ(C-OH) and δ (C-OH) at 1127 cm^{-1}; δ(C-OH) and δ(C-H) at 1384 cm^{-1}; and υ(C-H) at 2910 cm^{-1}, weaken during starch gelatinization as a result of molecular order loss. We did not find a strong correlation between starch crystallinities and areas of the υ(C-H) band at 2910 cm^{-1}, and this might be attributed to the complex nature of this band.

The bands at 1384 cm^{-1}, 1341 cm^{-1}, 1264 cm^{-1}, 1127 cm^{-1}, 1084 cm^{-1}, and 1053 cm^{-1} are dominated by C-OH modes (Table 2), and the OH groups might contribute to intra-molecular hydrogen bonding between polymer chains. Intra-chain interaction by hydrogen bonding is a characteristic feature of starch crystallization. Intensities of the bands at 1264 cm^{-1} and 1053 cm^{-1}, showed direct dependency on crystallinity. However, sensitivity to variations in degree of hydration, possible differences in strengths of intra-chain hydrogen bonding among the starches, or polymorphic differences (Cael et al., 1973) might explain the slightly lower correlation coefficients (Figure 6b) as compared to that of the skeletal mode band at 480 cm^{-1}. Correlation of crystallinities with areas of the δ(C-H) shoulder at 1403 cm^{-1} was also strikingly strong (Figure 6b) suggesting crystallinity induced decoupling of the C-H and C-OH motions responsible for the broader band at 1381 cm^{-1}. Areas of skeletal-mode bands at 620, 530, 440, and 410 cm^{-1}, also correlated strongly with sample crystallinities, which we attributed to glucopyranose ring stabilization in crystalline domains.

Conclusion

In this study, we evaluated correlations between features in Raman signals and crystallinity of some starch powders. We have shown that there exist strong linear correlations between crystallinity and the Raman signals. With appropriate standardization, integrated areas of Raman bands can be used to generate calibration functions for determination of actual crystallinity values. The isolated skeletal mode at 480 cm^{-1}, which is strong and fairly symmetrical, appears to be the best for this purpose. However, intensity of this band might be influenced by polymer chain length. This suggests that chain-length distribution or amylose and amylopectin content might affect the integrated intensity, and so a calibration function based on this band may be limited to starches of comparable chain-length structure. Intensities of other bands particularly involving skeletal modes are also crystallinity sensitive. It remains necessary to evaluate these modes to establish how they relate to crystallinity in specimens of relatively diverse polymer-chain characteristics.

Acknowledgements

C. M. is grateful to Deutscher Akademischer Austauschdienst (DAAD) for financial support. The authors thank Dr. Thomas Doert and Dr. Silvia Paasch, Department of Chemistry, Technische Universitaet Dresden Germany, for assistance with X-ray diffraction and nuclear magnetic resonance spectroscopy measurements.

References

Buleon, A., Veronese, G., & Putaux, J. L. (2007). Self-association and crystallization of amylose. *Australian Journal of Chemistry*, *60*, 706-718.

Bulkin, B. J., Kwak, Y., & Dea, I. C. M. (1987). Retrogradation kinetics of waxy-corn and potato starches - a rapid, Raman-spectroscopic study. *Carbohydrate Research*, *160*, 95-112.

Cael, J. J., Koenig, J. L., & Blackwel.J (1973). Infrared and Raman spectroscopy of carbohydrates .Part III. Raman-spectra of polymorphic forms of amylose. *Carbohydrate Research*, *29*, 123-134.

Cael, J. J., Koenig, J. L., & Blackwell, J. (1975). Infrared and Raman-spectroscopy of carbohydrates .Part VI. Normal coordinate analysis of V-amylose. *Biopolymers*, *14*, 1885-1903.

Cairns, P., Bogracheva, T. Y., Ring, S. G., Hadley, C. L., & Morris, V. J. (1997). Determination of the polymorphic composition of smooth pea starch. *Carbohydrate Polymers*, *32*, 275-282.

de Veij, M., Vandenabeele, P., De Beer, T., Remonc, J. P., & Moens, L. (2009). Reference database of Raman spectra of pharmaceutical excipients. *Journal of Raman Spectroscopy*, *40*, 297-307.

Dupuy, N., & Laureyns, J. (2002). Recognition of starches by Raman spectroscopy. *Carbohydrate Polymers*, *49*, 83-90.

Fechner, P. M., Wartewig, S., Kleinebudde, P., & Neubert, R. H. (2005). Studies of the retrogradation process for various starch gels using Raman spectroscopy. *Carbohydrate Research*, *340*, 2563-2568.

Gernat, C., Radosta, S., Anger, H., & Damaschun, G. (1993). Crystalline parts of three different conformations detected in native and enzymatically degraded starches. *Starch-Starke*, *45*, 309-314.

Gidley, M. J., & Bociek, S. M. (1988). ^{13}C CP/MAS NMR-studies of amylose inclusion complexes, cyclodextrins, and the amorphous phase of starch granules - relationships between glycosidic linkage conformation and solid-state ^{13}C chemical-shifts. *Journal of the American Chemical Society*, *110*, 3820-3829.

Hermans, P. H., & Weidinger, A. (1948). Quantitative X-ray investigations on the crystallinity of cellulose fibers - a background analysis. *Journal of Applied Physics*, *19*, 491-506.

Imberty, A., Buleon, A., Tran, V., & Perez, S. (1991). Recent advances in knowledge of starch structure. *Starch-Starke*, *43*, 375-384.

Kim, I. H., Yeh, A. I., Zhao, B. L., & Wang, S. S. (1989). Gelatinization kinetics of starch by using Raman-spectroscopy. *Biotechnology Progress*, *5*, 172-174.

Kizil, R., Irudayaraj, J., & Seetharaman, K. (2002). Characterization of irradiated starches by using FT-Raman and FTIR spectroscopy. *Journal of Agricultural and Food Chemistry*, *50*, 3912-3918.

Lopez-Rubio, A., Flanagan, B. M., Gilbert, E. P., & Gidley, M. J. (2008). A novel approach for calculating starch crystallinity and its correlation with double helix content: a combined XRD and NMR study. *Biopolymers*, *89*, 761-768.

Ma, C. Y., & Phillips, D. L. (2002). FT-Raman spectroscopy and its applications in cereal science. *Cereal Chemistry*, *79*, 171-177.

Mutungi, C., Onyango, C., Doert, T., Paasch, S., Thiele, S., Machill, S., Jaros, D., & Rohm, H. (2011). Long- and short-range structural changes of recrystallised cassava starch subjected to *in vitro* digestion. *Food Hydrocolloids*, *25*, 477-485.

Mutungi, C., Onyango, C., Jaros, D., Henle, T., & Rohm, H. (2009a). Determination of optimum conditions for enzymatic debranching of cassava starch and synthesis of resistant starch type III using central composite rotatable design. *Starch-Starke*, *61*, 367-376.

Mutungi, C., Rost, F., Onyango, C., Jaros, D., & Rohm, H. (2009b). Crystallinity, thermal and morphological characteristics of resistant starch type III produced by hydrothermal treatment of debranched cassava starch. *Starch-Starke*, *61*, 634-645.

Paris, M., Bizot, H., Emery, J., Buzare, J. Y., & Buleon, A. (1999). Crystallinity and structuring role of water in native and recrystallized starches by ^{13}C CP-MAS NMR spectroscopy 1: Spectral decomposition. *Carbohydrate Polymers*, *39*, 327-339.

Paris, M., Bizot, H., Emery, J., Buzare, J. Y., & Buleon, A. (2001). NMR local range investigations in amorphous starchy substrates I. Structural heterogeneity probed by ^{13}C CP-MAS NMR. *International Journal of Biological Macromolecules*, *29*, 127-136.

Passauer, L., Bender, H., & Fischer, S. (2010). Synthesis and characterisation of starch phosphates. *Carbohydrate Polymers*, *82*, 809-814.

Phillips, D. L., Xing, J., Chong, C. K., Liu, H. J., & Corke, H. (2000). Determination of the degree of succinylation in diverse modified starches by Raman spectroscopy. *Journal of Agricultural and Food Chemistry*, *48*, 5105-5108.

Pifferi, G., Santoro, P., & Pedrani, M. (1999). Quality and functionality of excipients. *Farmaco*, *54*, 1-14.

Pohu, A., Planchot, V., Putaux, J. L., Colonna, P., & Buleon, A. (2004). Split crystallization during debranching of maltodextrins at high concentration by isoamylase. *Biomacromolecules*, *5*, 1792-1798.

Sajilata, M. G., Singhal, R. S., & Kulkarni, P. R. (2006). Resistant starch - A review. *Comprehensive Reviews in Food Science and Food Safety*, *5*, 1-17.

Santha, N., Sudha, K. G., Vijayakumari, K. P., Nayar, V. U., & Moorthy, S. N. (1990). Raman and Infrared-spectra of starch samples of sweet-potato and cassava. *Proceedings of the Indian Academy of Sciences-Chemical Sciences*, *102*, 705-712.

Schuster, K. C., Ehmoser, H., Gapes, J. R., & Lendl, B. (2000). On-line FT-Raman spectroscopic monitoring of starch gelatinisation and enzyme catalysed starch hydrolysis. *Vibrational Spectroscopy*, *22*, 181-190.

Sekkal, M., Dincq, V., Legrand, P., & Huvenne, J. P. (1995). Investigation of the glycosidic linkages in several oligosaccharides using FT-IR and FT- Raman spectroscopies. *Journal of Molecular Structure*, *349*, 349-352.

Tan, I., Flanagan, B. M., Halley, P. J., Whittaker, A. K., & Gidley, M. J. (2007). A method for estimating the nature and relative proportions of amorphous, single, and double-helical components in starch granules by ^{13}C CP/MAS NMR. *Biomacromolecules*, 8, 885-891.

Tester, R. F., Karkalas, J., & Qi, X. (2004). Starch - composition, fine structure and architecture. *Journal of Cereal Science*, *39*, 151-165.

Thygesen, L. G., Lokke, M. M., Micklander, E., & Engelsen, S. B. (2003). Vibrational microspectroscopy of food. Raman vs. FT-IR. *Trends in Food Science & Technology*, *14*, 50-57.

van Soest, J. J. G., Tournois, H., deWit, D., & Vliegenthart, J. F. G. (1995). Short-range structure in (partially) crystalline potato starch determined with attenuated total reflectance Fourier-transform IR spectroscopy. *Carbohydrate Research*, *279*, 201-214.

Wellner, N., Georget, D. M. R., Parker, M. L., & Morris, V. J. (2011). In situ Raman microscopy of starch granule structures in wild type and *ae* mutant maize kernels. *Starch-Starke*, *63*, 128-138.

Table 1: XRD crystallinity (%), and relative proportions of structural phases (%) by ^{13}C CP/MAS NMR

Sample	XRD Crystallinity (Xc)		Structural phase		
	Xc_i [a]	Xc_{ii} [b]	Amorphous	Crystalline	Interfacial
AS	0	0	100	nd	nd
DS	19.0 ± 1.3[c]	33.4 ± 1.3	49.9 ± 1.5	30.6 ± 0.8	19.4 ± 0.8
ANN-DS	23.9 ± 0.2	40.9 ± 3.2	33.9 ± 1.6	41.7 ± 2.5	24.4 ± 3.4
TC-DS	36.3 ± 1.7	50.7 ± 3.1	28.3 ± 2.3	52.2 ± 1.5	19.5 ± 2.0
HMT-DS	41.7 ± 1.7	56.2 ± 1.4	20.4 ± 1.8	56.0 ± 2.2	23.6 ± 3.5
dANN-DS	21.8 ± 0.5	47.9 ± 1.2	nd	50.9 ± 4.0	48.7 ± 2.5
dTC-DS	30.8 ± 0.9	54.4 ± 1.1	nd	56.9 ± 3.1	43.1 ± 2.5
dHMT-DS	33.4 ± 1.4	58.2 ± 2.4	nd	61.3 ± 3.0	37.1 ± 3.5

AS: amorphous starch, DS: debranched starch, ANN-DS: annealed debranched starch, TC-DS: temperature-cycled debranched starch, HMT-DS: heat-moisture treated debranched starch. dANN-DS, dTC-DS and dHMT-DS are partially digested forms of ANN-DS, TC-DS and HMT-DS, respectively.

[a] diffractogram-minima baseline approach.

[b] amorphous-template baseline approach.

[c] means ± standard deviation of four independent fitting procedures.

nd: not determined.

Table 2: Positions of Raman bands and their assignments

Band (cm^{-1})	Assignment	Reference
2910	υ(C-H)	a-c
1460	δ_s(CH_2) twisting, C-H bending	a, b, d
1403 (shoulder)	δ(C-H) bending	j
1381	δ(C-OH), δ(CH) bending, δ(CH_2) scissoring	a-c
1341	δ(CH_2) twisting, δ(C-OH) bending	a, b, d, e
1305	δ(C-H)	a
1264	δ(CH_2), C-OH, CH_2OH (side chain) related mode	a, b, d
1205	δ(C-H),	a
1155 (shoulder)	υ_a(C1-O-C4') α-(1 → 4) glycosidic linkage	d
1127	υ(C-OH), δ(C-OH), υ(C-O),	a-d
1110 (shoulder)	C-C, C-O, C-H related modes	a, b
1084	δ(C-OH) bending	a-c
1053	υ(C-OH), δ(C-OH) modes, υ(C-C)	a-d
1003	CH_2 related mode	a, h
943	υ_s(C1-O-C4')	a-c, f
910 (shoulder)	δ(C-OH), δ(CH_2), δ(C1-H)	a-d
865	υ_s(C1-O-C5), δ(CH_2), δ(C-H)	a-c, f
770	υ(C-C) ring mode	a, e
721	υ(C-C) ring mode	a, b e
411, 443, 480, 530, 576, 618	υ(C-C-O), δ(C-C-O) pyranose ring skeletal modes	a-c, g

δ: mode involves change of bond angle (deformation), υ: mode involves change of bond length (stretching). Subscript "*s*": symmetrical, subscript "*a*": asymmetrical

Key to references: (a) Cael et al., 1973 .(b) Santha et al., 1990 (c) Schuster et al., 2000; (d) Sekkal et al., 1995; (e) Kizil et al., 2002 (f) Passauer et al., 2010 (g) de Veij et al., 2009.

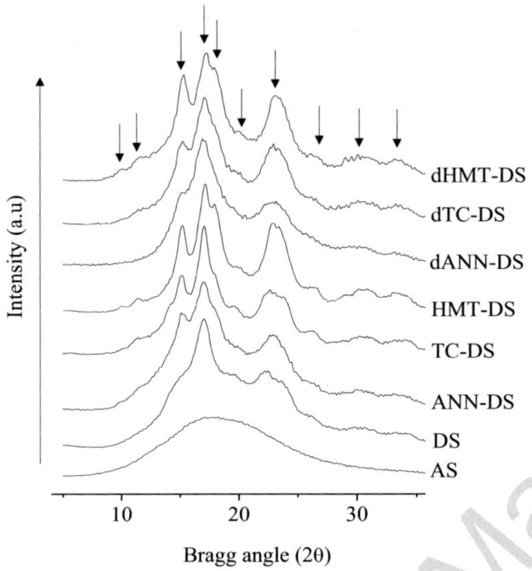

dHMT-DS
dTC-DS
dANN-DS
HMT-DS
TC-DS
ANN-DS
DS
AS
Intensity (a.u)
10
20
30
Bragg angle (2θ)

549

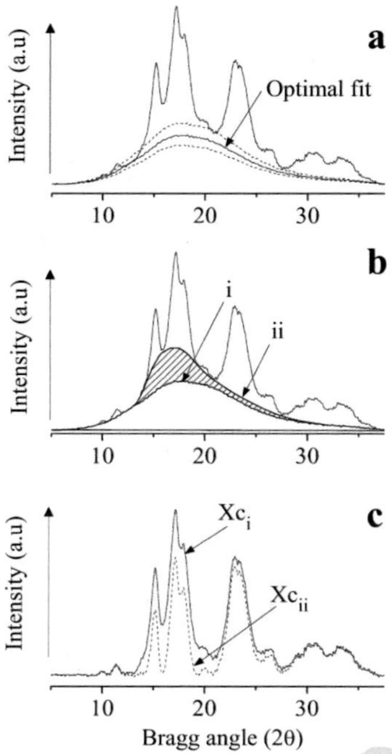
a
Intensity (a.u)
Optimal fit
10
20
30
b
Intensity (a.u)
i
ii
10
20
30
c
Intensity (a.u)
Xc_i
Xc_{ii}
10
20
30
Bragg angle (2θ)

550

551

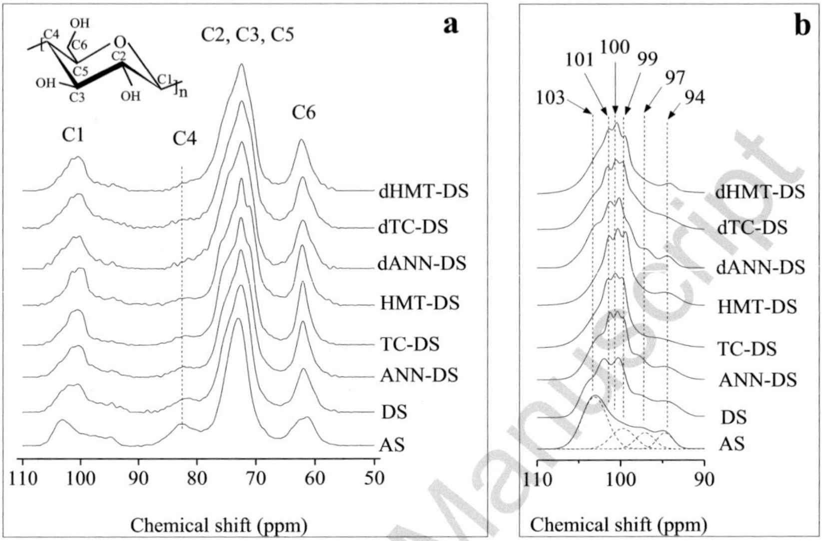
a
C2, C3, C5
C1
C4
C6
dHMT-DS
dTC-DS
dANN-DS
HMT-DS
TC-DS
ANN-DS
DS
AS
110 100 90 80 70 60 50
Chemical shift (ppm)
b
103
101
100
99
97
94
dHMT-DS
dTC-DS
dANN-DS
HMT-DS
TC-DS
ANN-DS
DS
AS
110 100 90
Chemical shift (ppm)

552

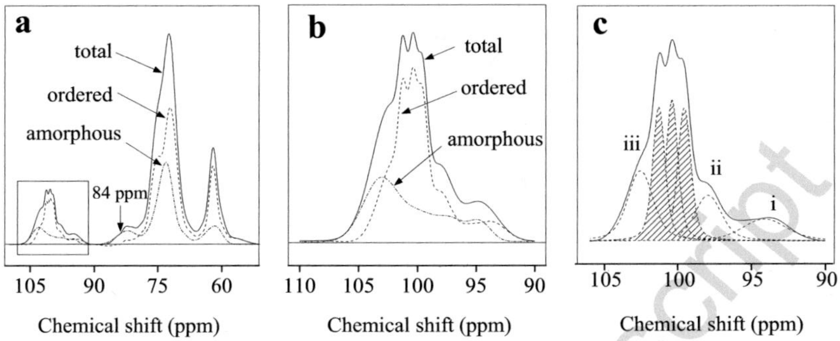
a
total
ordered
amorphous
84 ppm
105
90
75
60
Chemical shift (ppm)
b
total
ordered
amorphous
110
105
100
95
90
Chemical shift (ppm)
c
iii
ii
i
105
100
95
90
Chemical shift (ppm)

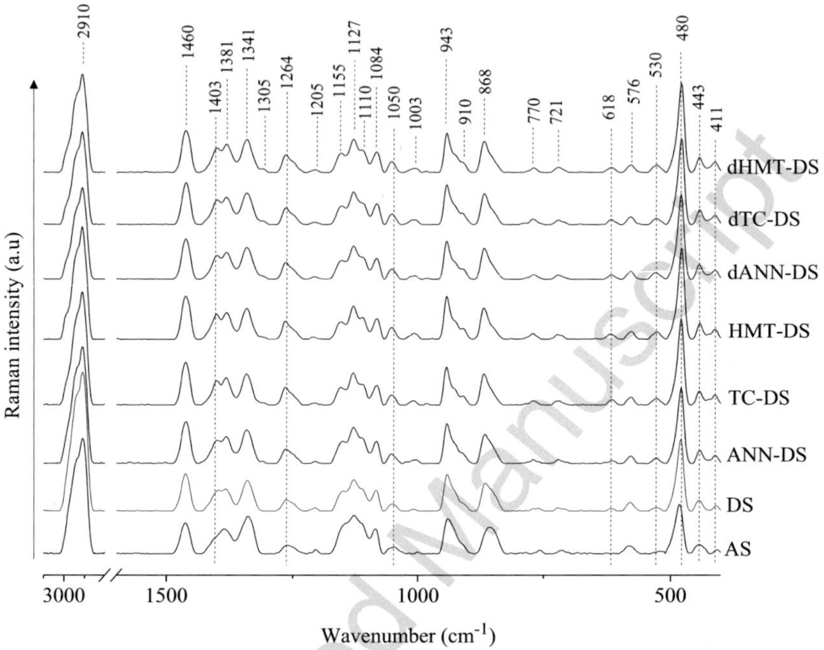
2910
1460
1403
1381
1341
1305
1264
1205
1155
1127
1110
1084
1050
1003
943
910
868
770
721
618
576
530
480
443
411
dHMT-DS
dTC-DS
dANN-DS
HMT-DS
TC-DS
ANN-DS
DS
AS
Raman intensity (a.u)
3000
1500
1000
500
Wavenumber (cm-1)

553

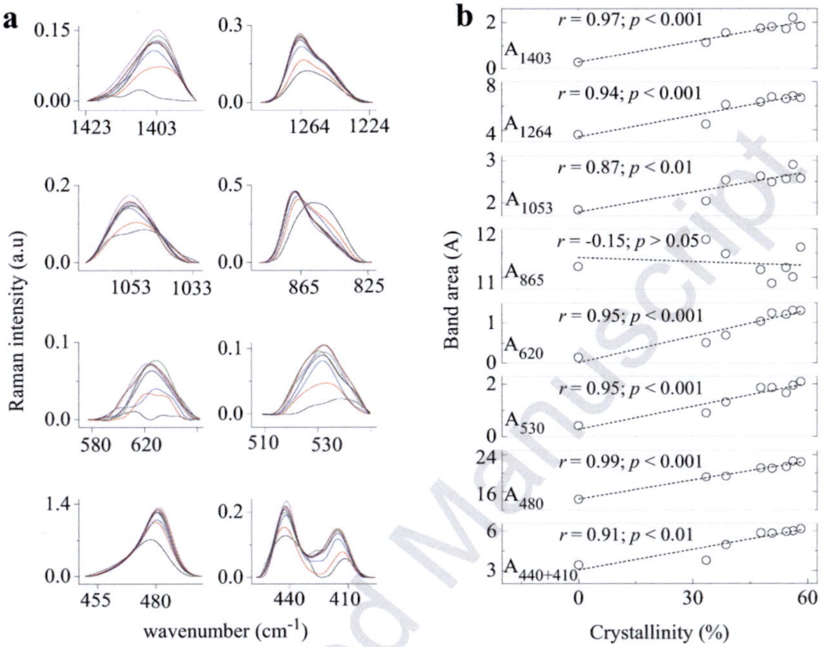
a
b
Raman intensity (a.u)
wavenumber (cm-1)
Band area (A)
Crystallinity (%)
A1403
r = 0.97; p < 0.001
A1264
r = 0.94; p < 0.001
A1053
r = 0.87; p < 0.01
A865
r = -0.15; p > 0.05
A620
r = 0.95; p < 0.001
A530
r = 0.95; p < 0.001
A480
r = 0.99; p < 0.001
A440+410
r = 0.91; p < 0.01

Figure 1: X-ray diffractograms of amorphous starch (AS), debranched starch (DS), annealed debranched starch (ANN-DS), temperature-cycled debranched starch (TC-DS), heat-moisture treated debranched starch (HMT-DS), and the partially digested forms: dANN-DS, dTC-DS and dHMT-DS.

Figure 2: Amorphous-template fitting on sample XRD diffractogram (a), decomposition of diffractogram by amorphous-template baseline [i] or by diffractogram-minima baseline [ii] fitting (b), and the difference spectra representing crystalline fractions, Xc_i and Xc_{ii}, by the two approaches, respectively (c). Shaded region in (b) represents diffuse scattering by imperfect and small crystallites.

Figure 3: ^{13}C CP/MAS NMR spectra of amorphous starch (AS), debranched starch (DS), annealed debranched starch (ANN-DS), temperature-cycled debranched starch (TC-DS), heat-moisture treated debranched starch (HMT-DS), and the partially digested forms: dANN-DS, dTC-DS and dHMT-DS. Broken line in (a) marks C4 atom signal associated with amorphous phase. (b) Deconvoluted profiles of the C1 atom signals; broken lines mark polymorph related peak multiplicities [101, 100, 99 ppm], and displacements arising from interfacial-phase contributions [103, 97, 94 ppm].

Figure 4: ^{13}C CP/MAS NMR spectrum decomposition into amorphous and ordered phases ((a) and (b)), and deconvolution of the ordered-phase C1 signal (c). Intensity of shaded peaks (triplet in A- and doublet in B-type starch) represent crystalline fraction. Intensities i, ii and iii are interfacial-phase contributions in the ordered-phase.

Figure 5: FT-Raman spectra of amorphous starch (AS), debranched starch (DS), annealed debranched starch (ANN-DS), temperature-cycled debranched starch (TC-DS), heat-moisture treated debranched starch (HMT-DS), and the digested forms: dANN-DS, dTC-DS and dHMT-DS. Broken lines cutting through all spectra mark bands whose intensities show marked crystallinity dependence.

Figure 6: Overlaid profiles of selected Raman bands of amorphous starch AS (black), debranched starch DS (red), annealed debranched starch ANN-DS (blue), temperature-cycled debranched starch TC-DS (green), heat-moisture treated debranched starch HMT-DS (magenta), and the partially digested forms: dANN-DS (navy), dTC-DS (violet), and dHMT-DS (brown) (a), and linear correlations between crystallinity and band areas (b). The 865 cm^{-1} band was used as internal standard and the band areas (A_{865}) are relatively invariable in the different samples.

Theses

1. Microbiota residing in the human gut is integral to health, and the growth, metabolism, and viability is dependent on the nature and supply of dietary components that escape digestion and absorption in the upper gut.

2. Some forms of starch resist digestion due to structural reasons, and so considerable amount of ingested starch reaches the lower gut where it is fermented into health promoting metabolites and helps in supporting a healthy gut microbiota balance.

3. In the literature, SCFAs, in particular butyrate are shown to mediate the majority of health benefits associated with resistant starch, and recrystallised starch (Resistant starch type 3, RS III) is a superior substrate for butyrate because of its structure that promotes saccharolytic microbiota and further tilts fermentation pathways in favour of butyrate production.

4. Resistance of recrystallised starch to enzyme digestion is due to realignment of linearised starch polymers into double helices, and the ordering or aggregation into supramolecular structures that hinder enzyme diffusivity, accessibility of glycosidic linkages, and effective binding.

5. Structure of recrystallised starch influences specificity of interaction with intestinal microbiota, and so structural features that influence enzyme attack are helpful for developing preparations that have specific functionality because these can be manipulated through processing methods to affect the nature, rate, site, and extent of colonic fermentation.

6. Delivery of the healthy benefits of RS III is possible through fortification of frequently consumed foods, and to enable enrichment in a wide array of foods, industrial-scale production can be made possible through starch conversion methods that yield preparations with a high content of the functional carbohydrate.

7. Recrystallisation of processed cassava starch into enzyme resistant form is enhanced by enzyme mediated hydrolysis of the α-D-(1 → 6) branching points because interaction between the starch polymers is enhanced when the latter

are linearised rather than branched.

8. The extent of enzyme-mediated removal of the branched polymers, alongside the structure of the resultant chains is dependent on the conditions of the catalytic transformation including the type, purity and activity of the debranching enzyme.

9. Formation of resistant starch is increased when low molecular weight products are separated from the linearised polymer chains by methods such as centrifugation, dialysis or filtration because they obstruct aggregation of the optimal chains.

10. Molecular structure of the base starch has additional influence on the structure, stability, as well as the physical properties of the recrystallised product such as solubility, water binding, swelling and matrix forming behaviour.

11. Treatment of debranched starch under controlled conditions of heat and moisture achieves enhancement of resistance to enzyme digestion and the techno-functional properties by way of increasing ordering and the density of helical packing of polymer chains in ordered domains. Such stabilisation is however partial and factors such as crystallite size, length of constituent polymer chains and the proximity of amorphous domains relative to the crystalline ones are bound to influence aspects related to state transitions and enzyme susceptibility.

12. Conformational order of the crystalline as well as the non-crystalline registry that characterise recrystallised starch influences interaction with other macromolecular species during processing and storage.

13. Both ordered and amorphous phases are integral part of recrystallised preparations and the ordered phase is further compartmentalised into crystalline double helices, non-crystalline double helices, constrained conformations and single helices depending on the method of recrystallisation.

14. Molecular order and crystallinity are important for resistance of recrystallised starch to digestion, but degree of structural heterogencity, that is, the relative proportions of amorphous conformations caged within imperfect crystallite,

non-crystalline double-helical structures, single helical chains, and constrained chain conformations also influence digestibility in a manner related to their proximity to the ordered or crystalline regions.

15. *In vitro* digestion of the recrystallised starch enhances crystallinity but structurally, crystallites are less rigid, smaller and probably distorted due disentangling effect of α-amylase. A-type crystalline packing decreases, the B- and V-type morphologies increase, and the disordered phases diminish.

16. Polymeric residues of *in vitro* starch digestion are highly ordered and the stoichiometry of constituent conformations is dependent on the structural rigidity of the starch before digestion.

17. Interaction with moisture is fundamental property of recrystallised starch and is key determinant of the interrelations with other ingredients in composite foods systems and stability against enzyme digestion.

18. In highly recrystallised starch moisture does not get absorbed into the matrices, and this makes it less competitive for moisture and therefore suitable in products where other macromolecular ingredients such as proteins, other carbohydrates or even salts need the moisture for attainment of specific handling or product qualities.

19. Inability of moisture diffuse into the internal matrices of highly recrystallised starch preparations explains in part the high resistance to digestion because enzyme and starch can not approach each other over fairly sufficient surface, and hydration constraints limit hydrolysis in the crystalline regions.

20. Modification of molecular structure of cassava starch prior to debranching, and subjection of the debranched starch to controlled thermal treatments in the presence of moisture provide some practical approaches to tailor-make the techno-functional properties of recrystallised starch to suit a wide range of food products.

Dresden, March 31, 2011 **Christopher Mutungi**